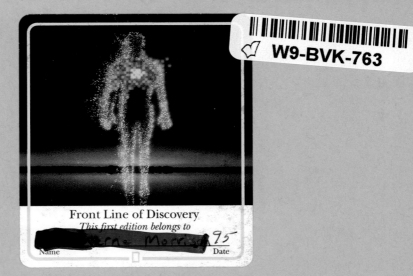

Front Line of Discovery
*This first edition belongs to*

Name ▮ Date

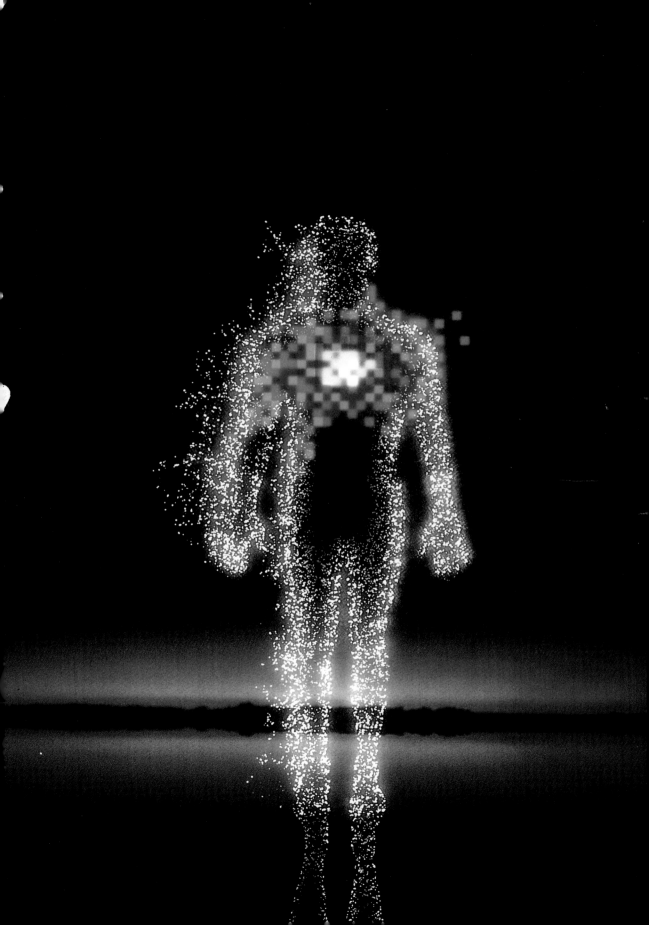

# FRONTLINE
## OF DISCOVERY

Science on the Brink of Tomorrow

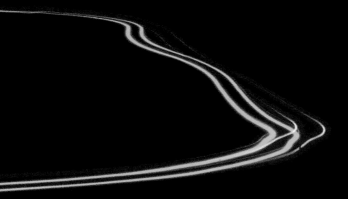

Prepared by the Book Division
National Geographic Society, Washington, D. C.

# FRONT LINE
## OF DISCOVERY

**Contributing Authors**
Arthur C. Clarke
Carole Douglis
Robert Friedel
Stephen S. Hall
Richard Restak
Dava Sobel
Walter Sullivan

**Contributing Photographer**
Roger H. Ressmeyer

**Published by**
The National Geographic
Society

Gilbert M. Grosvenor
*President and Chairman
of the Board*

Michela A. English
*Senior Vice President*

**Prepared by**
The Book Division

William R. Gray
*Vice President and Director*

Margery G. Dunn
Charles Kogod
*Assistant Directors*

**Staff for this book**
Barbara A. Payne
*Managing Editor*

Greta Arnold
*Illustrations Editor*

Marianne R. Koszorus
*Art Director*

Susan C. Eckert
Ann N. Kelsall
*Researchers*

Carole Douglis
Seymour L. Fishbein
Ann N. Kelsall
Tom Melham
Cynthia Russ Ramsay
*Picture Legend Writers*

Sandra F. Lotterman
*Editorial Assistant*

Karen Dufort Sligh
*Illustrations Assistant*

Lewis R. Bassford
Timothy H. Ewing
*Production Project
Managers*

Richard S. Wain
*Production*

Karen F. Edwards
Elizabeth G. Jevons
Peggy J. Oxford
*Staff Assistants*

**Manufacturing and
Quality Management**
George V. White
*Director*

John T. Dunn
*Associate Director*

George J. Zeller, Jr.
*Manager*

R. Gary Colbert
*Executive Assistant*

George I. Burneston III
*Indexing*

*Portfolio:*

*Page 1:* Composite
photograph of the
human form

*Pages 2-3:* Fiber-optic
cables, mainstay of the information superhighway

*Pages 4-5:* 700-pound
helmet of the gamma
knife, a new kind of bloodless brain surgery

*Pages 6-7:* Parabolic mirrors
focus the sun's energy to
convert it into electricity.

*Pages 8-9:* Remnant of an
exploding star illuminates a
gas cloud in the vast
reaches of the universe.

*Title page:* Time exposure
reveals Keck Telescope atop
Mauna Kea in Hawaii.

*Page 13,* top to bottom:
Patient undergoing computerized axial tomography.
Flasks containing cells in a
culture medium. Optical
fibers in an abstract pattern.
Fractal showing basic pattern of nature. Laser beam
aimed skyward at the
Starfire Optical Range.

Library of Congress CIP Data:
page 200

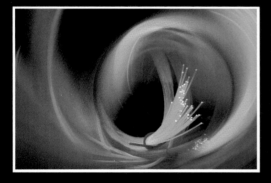

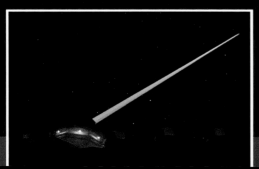

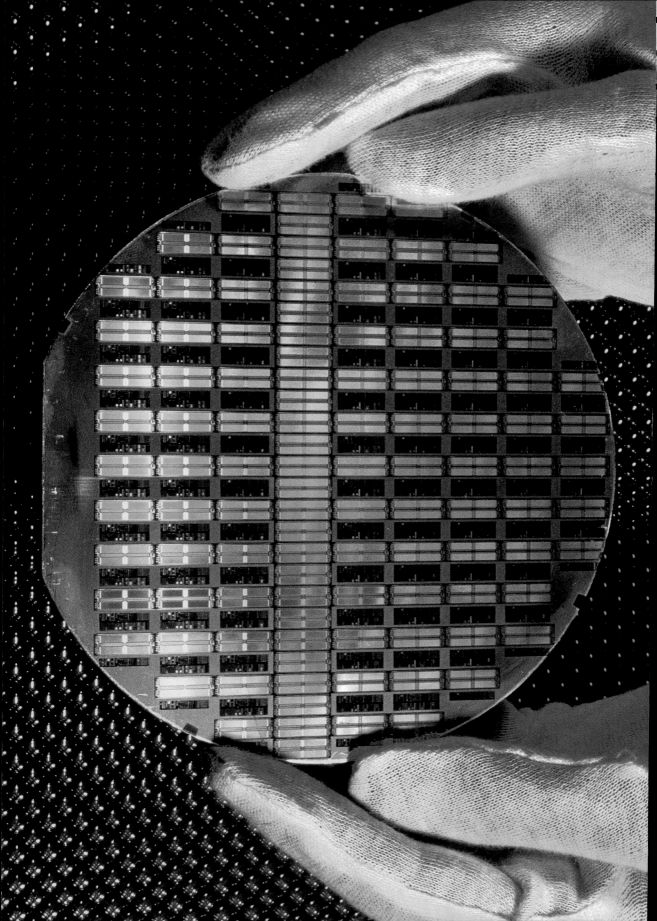

# FRONT LINE
# OF DISCOVERY

## An Introduction
## by Walter Sullivan

NEVER, IN ANY EARLIER CENTURY, has
there been such an avalanche of
revolutionary discoveries relating to life,
our planet, the nature of our universe,
and the material of which it is formed.
As we look back on those discoveries
we can guess some of the changes
that lie ahead, but there are bound
to be great surprises. Many of the most
dramatic developments of recent
decades have been during my own
career as a science watcher. They
include amazing findings on the deep
ocean floor made possible by special
submersibles, the creation of new elements, and the identification of new
components of the atom. Trains have

The integrated circuit, here a megabit chip, has revolutionized the world of science.

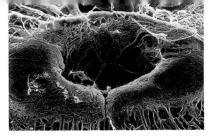

*Electron micrograph of a retina*

Telescope, have enabled us to see the universe with new eyes. Techniques developed for the Star Wars program have helped ground-based instruments observe as though they were above the atmosphere, eliminating the twinkle created as starlight passes through the air. Infrared observations have produced evidence that planets are forming around many young stars, and new observing methods may let us estimate more accurately how many of them resemble the earth. Some observatories are searching for signals indicating that intelligent, technological life has emerged on some of these worlds.

As new methods have opened our eyes on the most distant parts of the universe, so other computer-controlled machines have opened the door to nature's inner workings, giving us new glimpses into our own bodies and even into the inner components of the atom. Three-dimensional imaging techniques, radioactive tracers, and numerous other devices let us check the performance of heart, brain, and other organs.

To the ancient Greeks the word "atom" meant indivisible, but it has been learned, from giant atom-smashing machines, that the atom has many components, controlled by a variety of forces. The end is not in sight. The particle accelerations needed to dig even deeper require giant and very costly machines.

Almost as costly has been the research seeking to harness some variation of the sun's fusion process for energy production. The supplies of conventional fuel—coal, gas, and oil—are limited; and greater dependence on other forms of energy seems inevitable in the future. Hope rests in part on recent developments in solid-state physics, such as the production of substances that, with increasing efficiency, convert sunlight into electricity.

The most dramatic of these developments has been the discovery of materials that conduct electricity with no resistance. In 1911 it was found that when mercury is cooled to almost absolute zero—the absence of all heat—its resistance to electricity disappears. Other substances were found to superconduct in this way but at temperatures still so low only very specialized applications seemed likely.

Then, in 1986, two researchers found that a cooled compound lost all electrical resistance at a temperature higher than that of any other known superconductor. The discovery set in motion what the Royal Swedish Academy of Sciences, in awarding the Nobel Prize to the discoverers, called "an avalanche" of efforts to surpass their achievement.

The next year, the American Physical Society hastily organized a meeting to exchange reports on the new results. Seeing what was coming, I managed to sneak in a side door and get a seat. When the main doors opened there was a stampede. Some, recalling the famous rock concert in upstate New York, called it "the Woodstock of physics." In subsequent years, new superconducting compounds have been found.

———————

Today, we are in the midst of a revolution in communications, dependent on such competing methods as radio, fiber optics, voice and electronic mail, fax, and television. New ways to amplify light traveling through fibers has made it possible to transmit data across an ocean at 10 billion bits a second. Cellular phones and the like have to some extent made possible a wireless society. We can talk to one another even when en route by plane, train, car, or on foot. Our computers can converse with each other across oceans. Our electronics have replaced paper for data and history storage.

Modern science has confronted us with life-and-death choices that are new to human society. Laboratory techniques that enable otherwise infertile couples to bear children often result in multiple conceptions and selection of the "best" embryo, challenging our ethics. The same kind of choice confronts us in treating the elderly. With new medications, surgery, or devices, it is possible to keep alive those who otherwise would die, enabling them to survive for what is often a rich life but sometimes can be vegetative.

Our success in battling disease, prolonging life, and producing more babies has led to other problems. Our population is beginning to overwhelm the planet. We live closer and closer to our neighbors. Constantly expanding air travel and foreign trade may spread pests and diseases more rapidly than science can cope with them. Medicine races to try to keep ahead of such scourges as AIDS. Vaccines have done wonders in the past but the infecting agent may develop resistance, as with tuberculosis and malaria.

Never before have we had to adapt to such rapid and radical changes as those documented in the subsequent chapters. We live at a time of both great promise and great peril. We need a populace prepared to deal not only with new technologies and discoveries, but also with

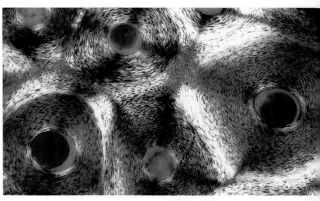

*Whale bone, magnified 50 times*

with the life-and-death decisions with which we are now confronted. The times call for wisdom and humanity, as well as skill and knowledge. It is a world full of uncertainties, but also vibrant with excitement.

# THE BRAIN

## by Richard Restak

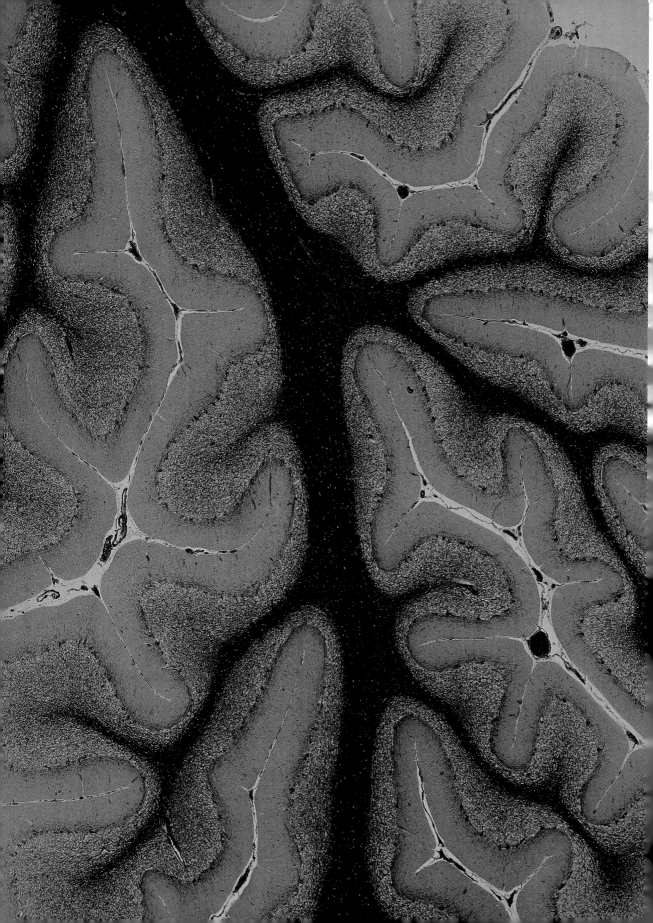

# THE BRAIN

## 1

IN AN OPERATING ROOM of the
Chicago Institute of Neurosurgery and
Neuroresearch, neurosurgeon Gail
Rosseau is moving a mechanical arm,
or viewing wand, over the surface of
her patient's scalp moments before
starting an operation to remove a brain
tumor. As she touches different points
with the wand, sensors at its tip convey
information about the patient's position
on the operating table and link the
information with a three-dimensional
picture on the computer screen. Within
minutes, Dr. Rosseau has before her
on the screen a three-dimensional
reconstruction of the patient's brain.

Cross-section of the human cerebellum, or small brain, reveals a leaflike pattern.
PRECEDING PAGES: A patient undergoes computerized axial tomography—a CAT scan.

**"Citadel of sense-perception,"** declared Pliny the Elder of the human brain. Myriad interactions of its billions of cells enable not only sensation but also movement, coordination, memory, judgment, and contemplation.

She manipulates the image into planes and sees structures she will encounter during the operation. Measurements are taken of volume, distance, and angle, and surgical cuts are simulated to reveal interior detail. The tumor is shown growing along the sphenoid, a bone that originates deep within the brain.

With a wave of the wand Rosseau, like some magic princess in a fairy tale, moves in and around and even through solid bone, tripping along the length of the sphenoid and exploring how the tumor's composition unfolds in relation to neighboring structures.

Just as she will do during the real operation, Rosseau cuts through the tumor on the screen and sees simulated in the computer image the results of her actions. She can even call on a "memory viewport" to compare this tumor with others she has operated on. Or a "text viewport" can be opened to provide textbook descriptions of variations of the tumor.

Dr. Rosseau sees on the screen the outline of the tumor and the overlying skin where she will make her incision. She is now intimately acquainted with the brain structures and their relationship to each other, even before she faces them deep within her patient's brain.

Devices such as the viewing wand, a creation of ISG Technologies in Toronto, will transform neurosurgery by enabling neurosurgeons to explore new operative approaches via interactive technology. This balances the physicians' desire for innovation with the ethical and humane demand that patients not be the subjects of experiments.

Technologies like the wand will answer the question: "What would be the result if I operated on this from another angle or approached the tumor from another plane of dissection?"

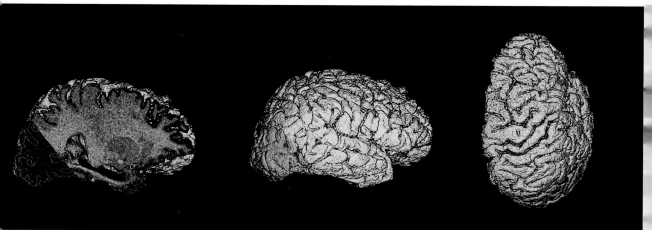

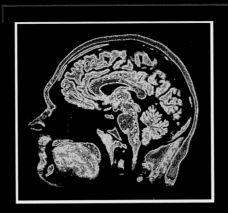

SAFELY ENCASED in a bowl of bone, the brain carries out its functions in physical seclusion from the rest of the body. The blood-brain barrier — a system of membranes that prevents substances from entering the brain — further protects this organ. Yet the brain's isolation is not total; nerve cells ferry electrochemical messages between it and other major sensory organs, as well as to receptors and muscles throughout the body.

In the coming decades, technology will continue to play a major role in opening what scientists traditionally thought of as a "black box": the human brain, which lies hidden within the skull and embedded within three enclosing membranes. This mysterious organ has fascinated and inspired researchers for as long as humans have wondered and speculated about their place in nature and the cosmos. It has intrigued me as a neurologist, serving as an impetus for applying new knowledge about it toward treatments for illnesses as varied as epilepsy, migraine headaches, and — as at the Chicago Institute — brain tumors.

Among the technological advances on the horizon are machines with interactive, three-dimensional capabilities.

Imagine a neurosurgeon sitting at home in his den the evening before a major operation. He is wearing a head-mounted display consisting of a pair of goggles through which he peers at a three-dimensional picture of the tumor

MOST COMPLEX ORGANISM IN THE UNIVERSE, the brain (below) mystified surgeons and scientists for centuries, confounding attempts to determine not only what it did, but how. Early anatomists dissected and named various brain structures; crude excisions of different parts determined which areas of the body they controlled. But just how the brain works remained largely unknown until the arrival of various computer-assisted techniques that permit physicians to "see" within living tissues.

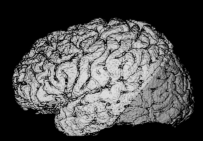
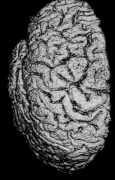
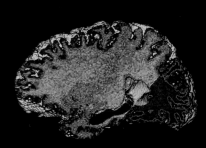

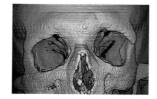

## 3-D BRAIN IMAGERY

New technologies help neurologists map the brain. While a CAT scan (left) relies on X rays, MRI—magnetic resonance imaging—uses radio waves. PET, or positron emission tomography, employs radioactive tracers.

he will encounter the next morning. Earlier, the neurosurgeon entered images of the patient's brain into a computer, which fashioned the information into the display he is now viewing through the goggles.

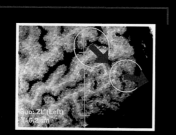

COMPUTER'S-EYE VIEW of part of a patient's brain shows a composite image assembled from different scanning techniques. Before the patient moves a finger, increased activity in the brain's hand-motor area (right arrow) indicates that neurons are sending a "move" command to the proper muscles. A few milliseconds later, cells in the hand-sensory area (left arrow) activate. The finger has moved.

After a few moments of study, the surgeon begins moving his instruments, which are attached to position sensors. As he manipulates the instruments, he sees the results of his actions on the patient's brain tumor. He is, in effect, rehearsing tomorrow's operation at home.

Significant reductions in surgical deaths and complications can be expected when neurosurgeons carry out such mock operations on preoperative, virtual-reality simulators. Like the flight simulators of the 1980s, they will allow physicians to perfect their skills and to transfer those skills to real-life situations. This will be especially valuable for inexperienced neurosurgeons or for those attempting new operations—

since, for a wide range of surgical procedures, errors are most likely during the first few operations performed.

But limitations exist for even the best imaging and the most realistic simulators. The brain doesn't entirely lend itself to scalpels and other traditional instruments of the neurosurgeon. Operative cures for brain tumors almost always involve tumors located around or under the brain. Diseases within the brain—for example, tumors that infiltrate the brain like ink soaking through a carpet—can rarely be totally removed because, like that ink-stained carpet, the tumor lacks a distinct margin, permeates widely, and requires removing large portions of brain to be certain of its complete removal. What's needed is something similar to a stain remover, which selectively attacks only those areas containing the ink stain while leaving the rest of the carpet undamaged.

Gliomas, the most deadly form of infiltrating brain tumors, kill 5,000 in the U. S. each year. The tumors are so lethal that fewer than 5 percent of the people diagnosed with them in 1994 will be alive in five years. Like Lee Atwater, the late Chairman of the Republican party, many victims die within a year. The failure of neurosurgical approaches for these malignant brain tumors is stimulating the development of a new era of bloodless brain surgery in which the steel scalpel is replaced with the gamma knife. What makes this such an astonishing breakthrough is that the gamma knife isn't a knife at all but an invisible blade of radiation beams.

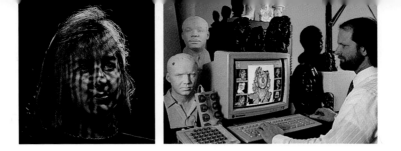

Created in Stockholm, the gamma knife consists of a large dome-shaped structure that emits more than 200 finely focused beams of radiation. After the patient's head is fixed within the unit, the beams of radiation simultaneously intersect at the precise location of the tumor. And since no incision is made—indeed, no operation in the traditional sense takes place—patients experience no discomfort, can move about later in the day, and may return home the next morning.

During the next several months after treatment, the tumor stops growing, reduces in size, and, if all goes well, may disappear altogether. In addition to helping patients suffering from brain tumors, the gamma knife has cured abnormalities

of the blood vessels that, left untreated, would surely have burst—killing or seriously crippling the patient. As the gamma knife becomes better known, the technique will be successfully applied to epilepsy and the elimination of small areas of the brain where seizures arise. It will also be applied to movement disorders such as Parkinson's disease, eliminating abnormal movements by applying ionizing radiation to specifically targeted brain centers.

A refinement of the gamma knife, the Acuray, under development at Stanford University, has an advantage: The patient's head need not be motionless. As it moves, so does the computer program controlling a robotic arm that delivers a fine beam of radiation to the target. The method relies on the same principle as a cruise missile: As the coordinates change, so does the program, which redirects the beam to the target's new position.

But the gamma knife and the Acuray, in common with standard neurosurgery, fail to influence the basic cellular dysfunction within brain cells that causes tumor cells to multiply. Molecular neurosurgery, presently an experimental procedure, holds promise for altering cellular

**CHECKERBOARD of light and shadow drapes a model (above left) as strobe lights facilitate computerized measurement of facial contours to make a 3-D image of her head. Sculptures of known dimensions (above right) help calibrate the scanner. Computers align scans into a single image (left) that reveals a brain tumor—all without making a single cut.**

**BLASTING EVEN A TINY TUMOR** demands massive hardware and software: Only a partition away from the operating room, a linear accelerator drives a rotating arm that delivers powerful electron beams to the target area. An alternative to the gamma, the "linac" requires more setup time but can access regions beyond the gamma knife's reach.

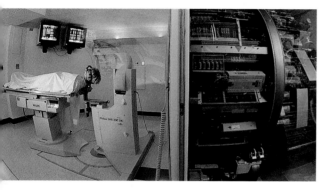

function through the use of a genetically engineered virus that homes in on tumor cells while sparing normal brain cells.

At Georgetown University Medical School in Washington, D. C., neurosurgeon Robert L. Martuza is using a one-step procedure aimed at destroying tumor cells. He has designed a genetically engineered virus that not only can kill malignant brain-tumor cells, but also is harmless to normal brain cells. When the virus is injected into the tumor, it spreads, killing any tumor cells it encounters. When the virus reaches the edge of the tumor, it stops growing.

"Our findings show that molecular neurosurgery can be used to kill intracranial tumor cells in a variety of nervous system tumors without damage to sur-

rounding normal brain cells," says Martuza.

A strategy that holds great promise involves administering a virus that makes it possible to selectively alter malfunctioning neurons, or brain cells, that are responsible for certain kinds of epilepsy.

"Traditional treatment for epilepsy involves patients taking medicines that bathe the entire brain, despite the fact that only a small number of malfunctioning cells may be responsible for the disorder. With molecular neurosurgery, we may be able to insert a gene that affects only malfunctioning cells by modulating the action of one neurotransmitter (chemical messenger) or enhancing the action of another," says Martuza.

"Over the next decade, look for current techniques of genetic engineering to be developed to create viruses with specific strategies in mind," he adds.

The most revolutionary advances in our understanding of the brain during the next decades will most likely come from new discoveries and refinements in our understanding of the brain's molecular biology. Over the next 15 years, scientists will try to map the human genome—an estimated 100,000 human genes that spell out the entire message conveyed by three billion letters of deoxyribonucleic acid, or DNA. This will be of significance to brain research because many brain diseases are known

## THE GAMMA KNIFE

**Using beams of radiation instead of surgical steel, this Swedish breakthrough "cuts out" a tumor without drawing blood. Doctors (left) examine brain scans on a light table to determine the target area.**

to be inherited. But knowing the genetic basis for a brain disorder may still leave us without a means of curing or even significantly influencing the disease.

Consider Tyron, a nine-year-old, mentally retarded boy who has to be restrained to prevent him from committing acts of self-mutilation. He has bitten off the ends of his fingers and has inflicted such damage to his lips that doctors are considering removing his teeth. The illness responsible for Tyron's behavior is called Lesch-Nyhan syndrome. Until recently the cause of this bizarre syndrome was a mystery. But scientists have now identified the defective gene. The protein produced by this gene has also been isolated. Yet despite detailed genetic and molecular information, doctors have no idea why patients like Tyron *(continued on page 34)*

**700-POUND HELMET** of the gamma knife immobilizes the patient's head while it aims beams of radiation precisely at a tumor or other structure. Although this and more conventional techniques of brain surgery can successfully eliminate some tumors, they do nothing on the genetic or cellular level to prevent them from forming or reforming.

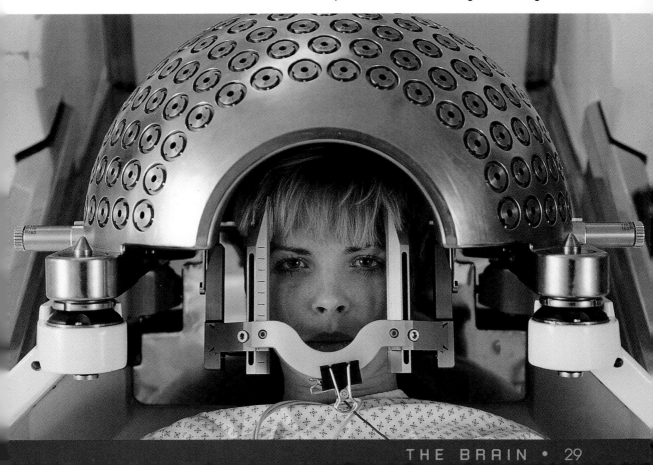

# BRAIN SURGERY

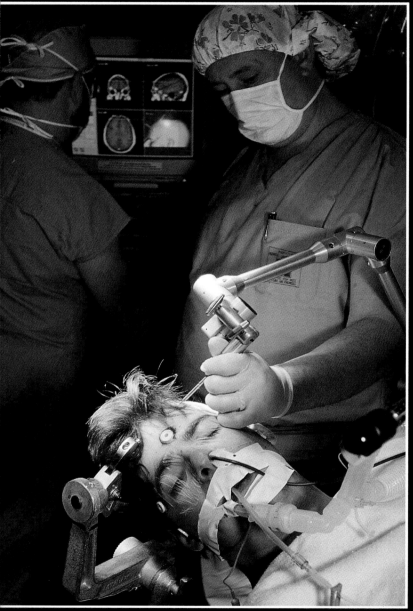

*The viewing wand helps doctors plot the surgical route to a tumor, by simulating views of the patient's brain as the wand moves over the scalp.*

**A**s recently as the mid-1960s, few could say, "I survived a brain tumor." Terror and helplessness surrounded those afflicted with malignancies in the brain, as little could be done. Today, while no patient looks forward to brain surgery, a large measure of hope now buffers the fear; a technological revolution affecting every aspect of the operating room has helped make the removal of brain tumors not only possible but routine. At the Chicago Institute of Neurosurgery and Neuroresearch, for example, various computerized imaging techniques enable surgical teams to diagnose a problem, "practice" different surgical solutions, and ultimately perform the operation along the best possible route. One system, Allegro, uses a jointed viewing wand (left) equipped with sensors to correlate its position on the patient's skull with recent CAT scans; a

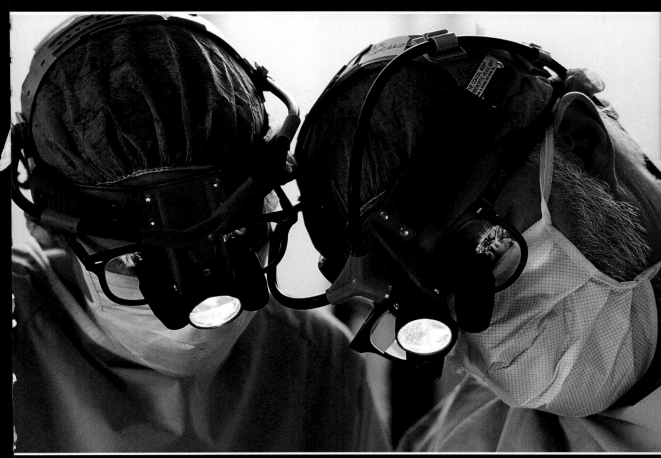

*Tools of the trade: high-power lenses and fiber-optic lamps*

computer assembles two- and three-dimensional views of the patient's brain on a monitor, enabling doctors to "see" the tumor and other structural details. Potential incisions can be simulated, to gauge their conse- quences before they are actually performed. A his- tory of similar tumor types and which operative pro- cedures were used also can be called up from the machine's memory. In such ways, the surgeon gets to "know" each brain and tumor before making the initial cut.

FOLLOWING PAGES: Filling a light box in the operat- ing room, CAT and MRI scans of the patient offer surgeons instant refer- ence points.

*Suction and retractor help extract a glioma tumor.*

*Malignant tumor*

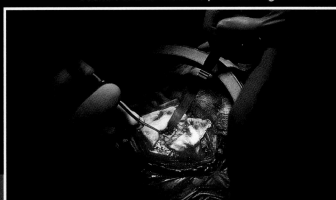

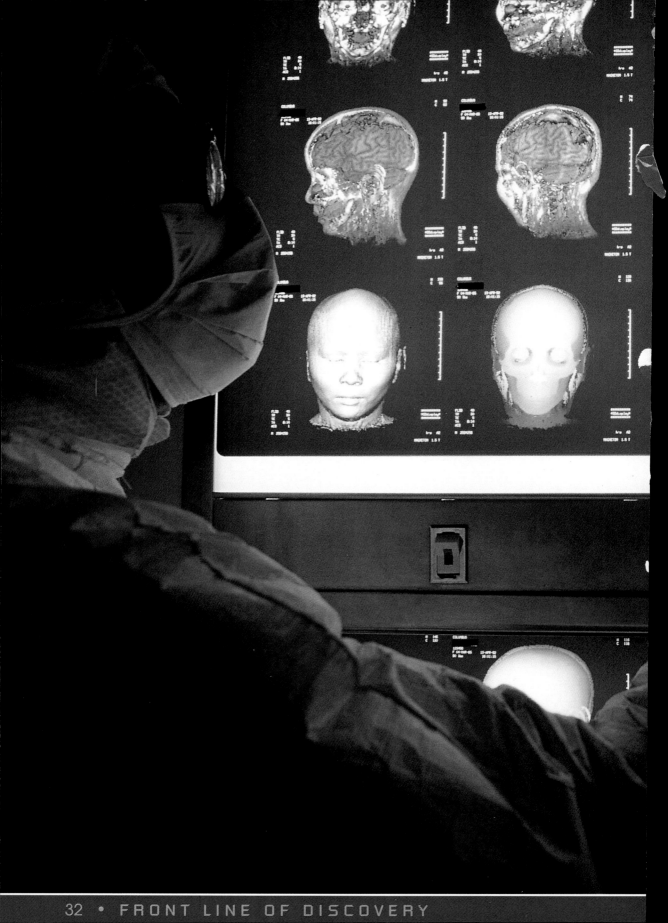

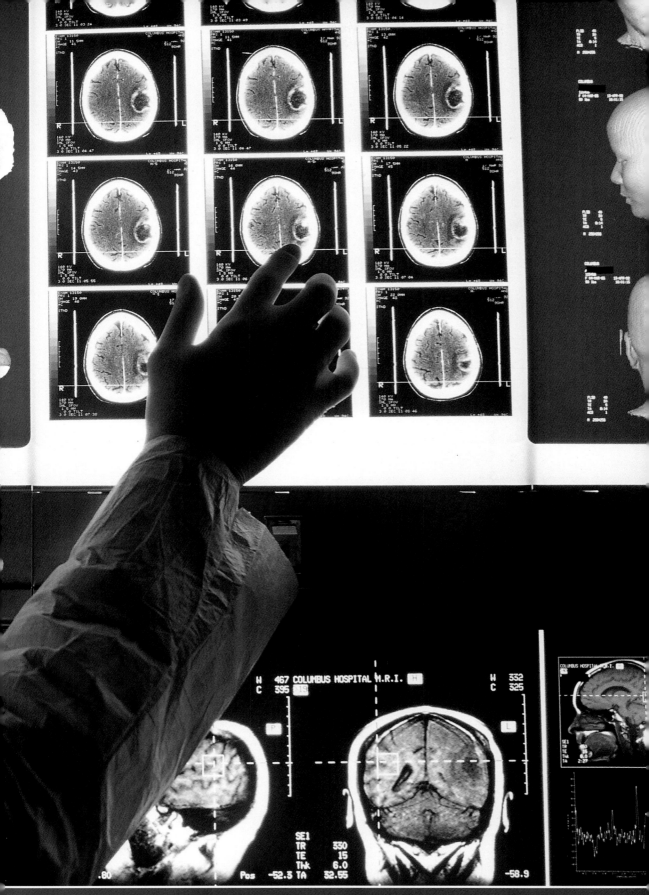

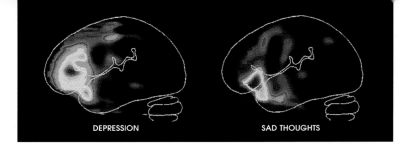

DEPRESSION          SAD THOUGHTS

mutilate themselves. Genes have to be understood in the context of brain structure and function. And at the moment the majority of instances of self-mutilation and mental retardation remain mysterious.

But neuroscientists are making progress in unraveling the interrelations of

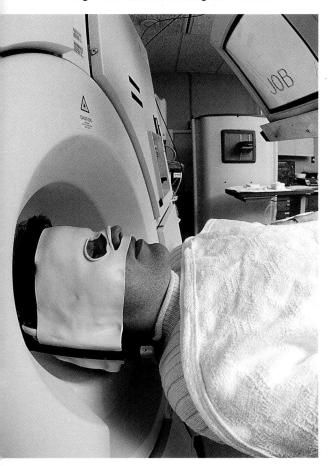

**THE WORD "JOB" serves as the stimulus for a volunteer patient during PET scan research. This study seeks to determine the location of the brain's seat of language—both for viewing words and for reading them aloud.**

genes to other neuropsychiatric disorders. Genetic discoveries over the next decade will change our ideas about disorders of mood (depression and mania), anxiety (panic, phobia, obsession-compulsion), and thought (schizophrenia)—the three major classes of inherited mental illness. Rather than single diseases, each category will likely consist of genetically distinct but overlapping disorders. Moreover, each variation of these illnesses is probably caused by distinct genetic mechanisms.

Manic depression, also known as bipolar disorder, is already recognized as resulting from genetic heterogeneity— abnormalities in several genes rather than a single gene. Different forms of the illness result from mutations in different genes in different affected families. Schizophrenia, too, is likely to turn out to comprise many genetically based variations of illness, sharing some features but differing in others.

Presently, more than 40 disorders of the brain, spinal cord, and peripheral nervous system are associated with gene mutations. Huntington's disease can now be diagnosed before birth by testing fetal cells obtained at amniocentesis. But since it is still untreatable, the use of tests for anything other than prenatal screening raises delicate ethical questions. Who has a right to know whether a person carries the gene for Huntington's?

Such decisions are complicated by the fact that genetically based brain diseases differ markedly in their expression. In instances of complete penetration of the disorder, `everybody who's got the

New imaging techniques allow neurologists to locate where thoughts occur, as well as discover how the brain directs functions. The scans at left represent a group of depressed patients and people thinking sad thoughts.

gene is going to have the disease. And they're going to have it in a devastating fashion," says Albert R. Jonsen, Chairman of the Department of Medical History and Genetics at the University of Washington.

But such situations are relatively rare when it comes to genetic diseases, which usually vary in their expression. One person may inherit the most serious form of an illness while another may inherit a form so mild that he or she has no inkling of its presence. Thus, the genetic diagnosis of brain disease is likely to remain of only limited value since tests available in the foreseeable future will measure a person's susceptibility to a disease rather than indicate the seriousness of the disorder.

Ethical challenges will arise from changes in our ideas about what constitutes a disease as opposed to normal human behavior. We are already witnessing this in regard to less severe behavioral dysfunctions. One of my own patients, a Washington, D. C., real estate broker I'll refer to as Peter (not his real name), has always found it difficult to keep himself motivated. Sometimes he didn't want to get out of bed and go to work, and generally he exhibited a low energy level — traits that over a five-year period cost him jobs and promotions. But Peter no longer experiences these difficulties since starting on one of a new generation of antidepressants, which work by altering the balance of the chemical messenger serotonin. While taking the drug, Peter no longer has difficulty motivating himself. He enthusiastically (continued on page 40)

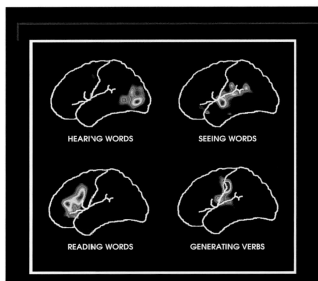

HEARING WORDS          SEEING WORDS

READING WORDS          GENERATING VERBS

COMPREHENDING THE MIND: PET scans of the brain's left hemisphere show strikingly different patterns of cerebral activity during four different aspects of a single function, language. The most active areas register red, while progressively less active regions show up as orange, yellow, green, blue, and black. Such techniques help determine not only which parts of the brain direct particular activities, but also which areas rule various emotions. Many researchers envision a massive "brain mapping" project that eventually would chart pathways for all cerebral functions. In scope it promises to be far more complex and challenging than current efforts to map the location of all human genes. But once completed, say proponents, such an "atlas of the mind" should lead to advances in mental health therapy.

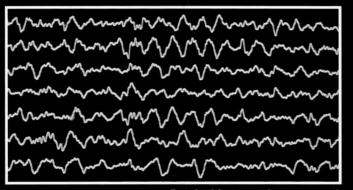

*Readout from a polysomnograph*

*Polysomnograph electrodes attach to head, chest, and face.*

**W**hat is this thing called sleep? Though we speak of snoozing "like a baby," or "like a log," few of us actually do. Sleep, researchers find, is not a passive phenomenon during which the brain simply rests—as once thought—but a state actively generated by an undetermined number of sleep centers in the brain. Jagged squiggles of a polysomnogram show electrical activity of the brain, heart, and facial muscles during sleep (upper left). Controversial research known as "lucid dreaming" uses flashing red lights to make dreamers

**SLEEP'S DIVERSE** states of mind show up as vividly colored patterns on PET scans, which image the living brain by using radioactive tracers to measure different metabolic rates and blood-flow shifts that occur during various mental activities. Stages shown depict (from left) wakefulness, normal sleep, severe sleep deprivation, REM sleep, and non-REM sleep. Colors show the measure of mental activity at the different stages. The more active colors of red and orange appear during periods of wakefulness and REM, when dreaming occurs.

**AWAKE**

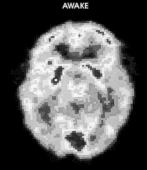

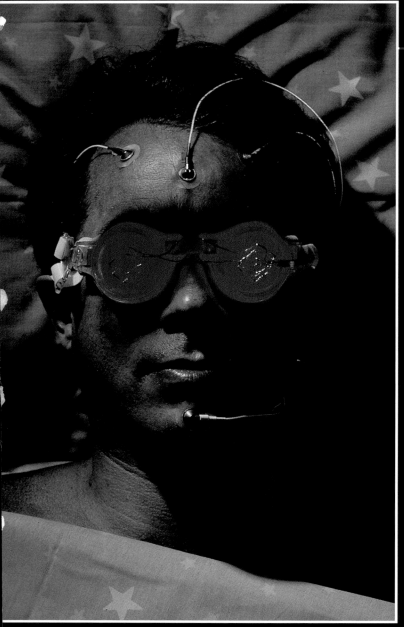

aware their visions are dreams even as they take place (left). In 1951, the discovery of REM sleep—short for rapid eye movement, which follows periods of deepest sleep and is marked by the most vivid dreams—added further evidence that the so-called sleeping brain actually is a hive of activity. During REM, pulse, blood pressure, and breathing become irregular. After about 20 minutes of REM, the sleeper drifts back into non-REM sleep in a cycle that repeats throughout the night.

FOLLOWING PAGES: Normal sleepers—even those who say they never move—average 8 to 12 posture shifts a night.

*Sensors and pulsing lights: the hardware of "lucid dreaming"*

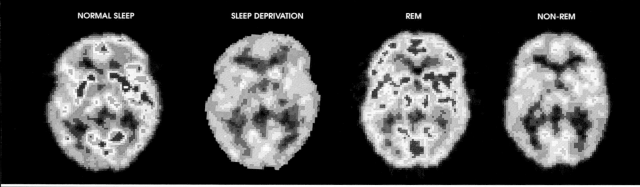

| NORMAL SLEEP | SLEEP DEPRIVATION | REM | NON-REM |

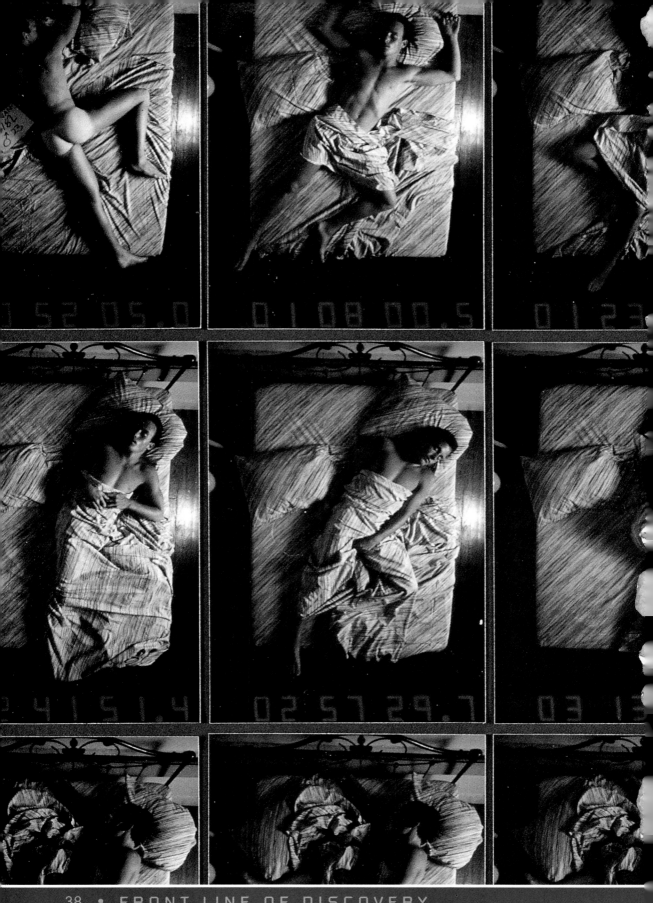

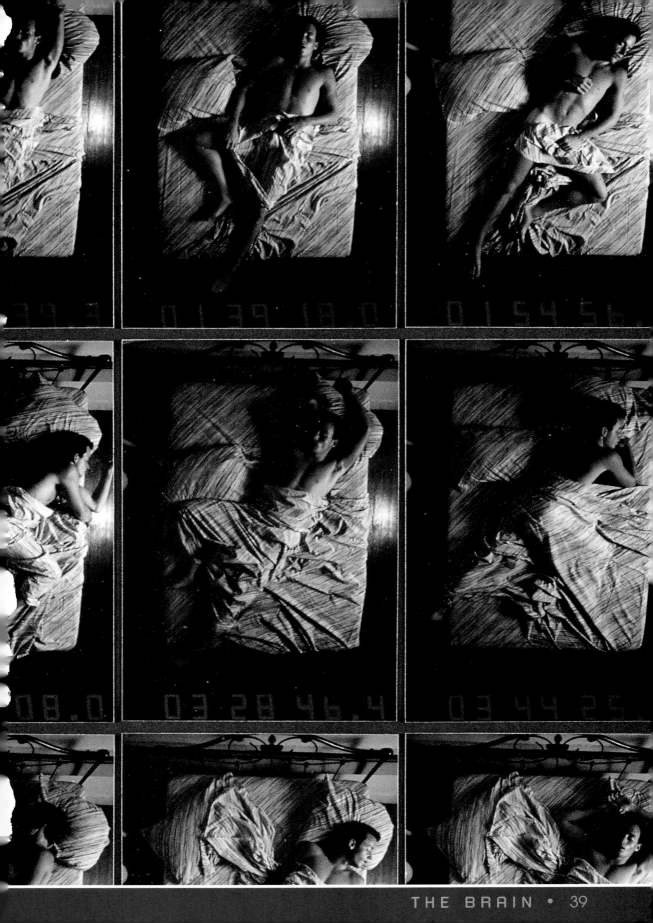

**GOGGLES WITH PULSING LIGHTS** accompany the serene surroundings of a Tokyo "mind gym," one way this congested city copes with stress. Like meditation, such gyms seek to arouse the brain's "relaxation response" by freeing the mind from day-to-day cares.

speaks of it to anyone who will listen and proclaims, "I want to stay on this drug for the rest of my life."

At the moment more than 15 million people like Peter worldwide are taking drugs affecting serotonin. Many suffer from severe depressions, and for them the drugs have been a lifesaver. But how many others are like Peter, either not suffering from depression at all or suffering from mild forms of the disorder that don't require long stretches of time on the drug?

Problems in living and unhappiness with one's life are increasingly being translated into some form of emotional illness, usually depression. Since additional antidepressants will soon be on the market, many with few side effects, it is likely that more people with mild or even nonexistent depression will be on mood-altering drugs. This not only reflects a change in traditional ways of thinking about mental and emotional disease; it also carries troubling implications about our national character.

What does it say about us when at a dinner attended by a *New York Times* reporter, four out of six highly successful professionals in their mid-30s, who agreed they "weren't seriously depressed anymore," revealed they were taking the antidepressant Prozac?

Less controversial will be the development of new drugs to combat drug addiction. A deep alteration of the brain is common to all forms of addiction, even legal substances like alcohol, nicotine, and caffeine. Deep-lying brain structures, with names like the nucleus accumbens and the septal nucleus, compose a "pleasure center" that is activated whenever we feel good. Addicting drugs act here too, and in the absence of these substances, the "crash" ensues—one of the reasons drug habits are so hard to break.

Within the next decade, drugs aimed at blocking the effects of addicting drugs, lessening craving, and reducing withdrawal effects will be developed.

In small doses, pain benefits us; it is the body's way of telling the brain that something is wrong. However, many threats to life can be painless, while others debilitate by overstimulating the brain's pain centers.

Milder addicting substances will substitute for stronger ones. And even more novel approaches are likely—such as altering the blood-brain barrier to prevent addicting substances from getting from the bloodstream into the brain.

Since we possess at birth all the neurons we will ever have, deafness, blindness, and paralysis exert catastrophic effects. Treatment for these conditions traditionally involves substitution rather than correction. A teletype substitutes for the telephone; Braille provides an alternative to sight; a wheelchair takes over the job formerly carried out by the legs. But recent advances in electronic engineering, especially the development of miniature electronic implants, promise corrective rather than substitutive aids. The impetus for this advance has come from new insights into how the brain works, combined with engineering advances that have made brain-machine couplings a reality.

Neural-prosthetic devices work by exchanging information directly between electronic devices and neurons. They are possible because of the similarities in information transmittal within the brain and the computer-assisted devices. Successful applications involve neurons conveying control signals to the prosthetic device, which then sends sensory information or motor commands directly to the brain.

FACES OF PAIN stem from causes as varied as the shock of a newborn's first bath or the prolonged stress of mental depression. Current treatments are equally diverse, ranging from pharmacological painkillers to acupuncture (below). Research has shown that the brain can mediate or even eliminate this most subjective of sensations by generating controlling chemicals that include numerous natural opiates, some far more powerful than morphine.

DIVERSE DRUGS—from nicotine in tobacco (above) to heroin (right)—share a potential for being addictive. Research indicates that centers for pleasure and pain exist in the brain's emotion-oriented areas. Various chemicals stimulate these centers, often by altering neurotransmitters—naturally occurring substances that carry messages between neurons.

For instance, deafness that is caused by illnesses affecting the connections between the ear and the brain frequently results from loss of the cochlear hair cells. These cells within the inner ear transform the wave vibrations of sound into nerve impulses. If they are damaged, a cochlear prosthesis can restore the perception of sound by stimulating the auditory nerve

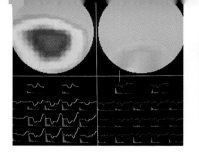

directly with electrical patterns delivered by a tiny array of electrodes surgically implanted into the inner ear.

Another neural-control device predicted to be available eventually is an electronic eye, or visual prosthesis. Already in place are functional neuromuscular stimulators (FNS) that enable paraplegics to walk. We've known since the 17th century that electrical pulses can animate a paralyzed leg. Even in cases of total paralysis, the nerves and muscles of paralyzed limbs respond to electrical stimulation. FNS takes advantage of this functional capacity by using implanted stimulators to command and control the paralyzed limbs.

At the moment, the field of neural prosthetics is one of the most promising and exciting areas of brain research. Dr. Richard Ruopp has learned this firsthand.

In 1988 the retired college vice president began experiencing slurred speech. Three months later the speech was worse, accompanied by twitching muscles in the right hand, foot, and lower lip. By the time of his diagnosis, Lou Gehrig's disease, or amyotrophic lateral sclerosis (ALS), Ruopp could barely lift his left foot. Walking was almost impossible, and a respirator became necessary. Rather than give in to the disease, the computer operator and teacher developed a computer-based communication system using a voice synthesizer of his own creation.

Even though his illness has progressed—ALS remains untreatable—Ruopp has used his prosthetic device to continue a lecture career, co-author a sci-

ence education textbook, and in 1991 marry his wife, Pat. He conveyed his "I do" through the computer-voice synthesizer.

Looming on the horizon are telecommunication programs capable of converting written text into speech for the blind and speech into text for the deaf. Speech recognition devices will allow a disabled person to operate a phone by voice rather than by touch. As Bishnu Atal of AT&T's Research Department put it: "My feeling is that within five years you will see dialogues in which a person can talk to a machine the way I am talking to you."

Those who stand to gain the most

**HIGH ON COCAINE, a laboratory rat exhibits humanlike addictions though its brain is less advanced. Given the chance, it chooses the drug over food, even when doing so involves pain or death by starvation. Like human addicts, rats show no satiation level for many drugs.**

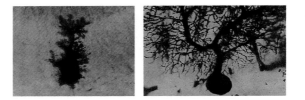

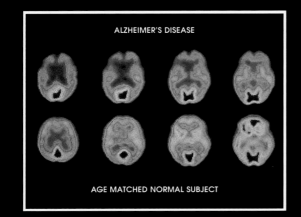

ALZHEIMER'S DISEASE

AGE MATCHED NORMAL SUBJECT

PET SCANS of normal brains and those afflicted by Alzheimer's disease show characteristic differences in areas and levels of activity (left), while autopsies reveal sharp structural contrasts in brain tissues. Well-formed crenellations and lobes define the normal brain (lower left); the diseased one (lower right) exhibits degeneration, having lost up to a third of its original bulk.

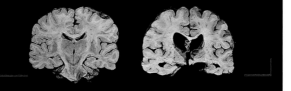

VICTIMS OF DOWN'S SYNDROME, a genetic disorder related to the presence of three copies of chromosome 21 rather than the normal pair, share some structural similarities with Alzheimer's patients—including this woman (opposite), once a nurse. Thus, researchers see this chromosome as a possible location for genes that can trigger— or prevent—Alzheimer's disease.

from prosthetic devices are the young— frequently victims of accidents. Their intellectual powers are unaffected, and they possess a high degree of personal motivation. In short, neural prosthetics are marvelous examples of emerging and revolutionary applications of brain research.

"Neural prosthetics today are at a level of development comparable to cardiology 25 years ago, when pacemakers were novel and primitive, and artificial hearts were a dream. The nervous system is certainly more complicated than the heart, but our technology is now vastly more sophisticated. The next 25 years will be seminal for applied neuroscience," according to Gerald E. Loeb of Queen's University in Kingston, Ontario.

Research on the normal brain has always taken second place to the study of disease. Most techniques for studying the

Comparison of brain cells, or neurons, in a newborn and a one-year-old (left) reveals rapid physical development. With age, the number of neurons erodes continuously, resulting in gradual loss of brain weight.

brain have involved a degree of risk that could only be justified by the prospect of curing disease. New imaging techniques, which are safer and more appropriate for studying the brain, are helping to increase our understanding of the living organ.

In Dr. Marcus Raichle's lab at Washington University School of Medicine, St. Louis, a volunteer is prepared for a PET scan, short for positron emission tomography. A cyclotron prepares water laced with a trace amount of radioactive oxygen. The water is injected into a vein in the volunteer's arm. It flows via the bloodstream to the brain. The brain is imaged by the scanner, which records the radiation emitted from the water and transforms this information into visual displays of different brain structures. The more active the brain area, the greater the water accumulation due to the increased blood flow associated with brain activities. The images of functional brain activity appear as cross-sectional tomographic pictures on the video screen of the PET scanner.

After the initial injection, Raichle shows the volunteer a series of single words on a computer at the rate of one a second. A computer image is made of the blood-flow pattern that accompanies the viewing of the words. After another injection, the volunteer reads aloud words that appear on the screen. Seconds later the computer reconstructs the new blood-flow patterns corresponding to reading words rather than merely looking at them.

"Now by subtracting blood-flow measurements made in a control state (viewing the words) from those in a task state (speaking the words), it is possible to identify those areas of the brain concerned with mental operations unique to the task of reading words," says Raichle. Called image *(continued on page 48)*

*(continued on page 48)*

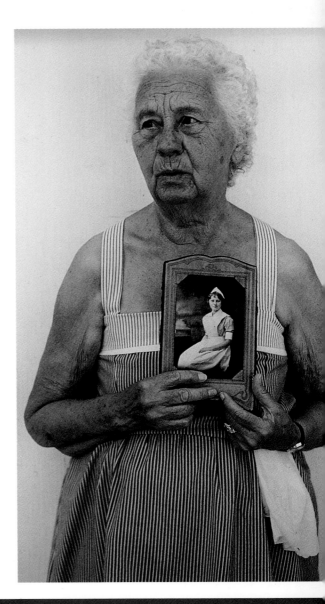

# NEUROPROSTHETICS

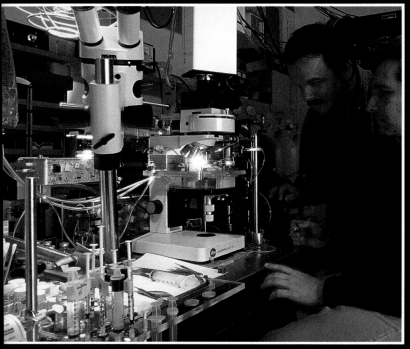

Researchers experiment with nerve endings and muscle control.

stimulated to walk again. What makes this possible is better understanding, not only of how the brain works but also of how normal neurons and muscles (opposite) interact. As our

Fingers "read" Braille.

knowledge—and technological ability—steadily improve, we find ourselves at the brink of realizing cures that for ages were considered miracles. Among them: empowering a victim paralyzed by nerve damage to regain the muscle control, coordination, and balance of a three-year-old ballerina (below).

U nlike most cells, neurons neither multiply nor regenerate; we possess our full complement at birth. Thus, damage to them can result in permanent effects such as blindness, deafness, or loss of motor control. Past treatments for such afflictions have centered on substitution: The blind use their fingertips to "read" Braille, while paraplegics

turn to wheelchairs for locomotion. But the recent emergence of neural prosthetics— devices that exchange information directly with neurons—promises new alternatives, often corrective rather than merely substitutive. Soon, the sightless may regain some vision through electronic eyes that communicate directly with the person's brain; paralyzed limbs can be electronically

Images of mouse tissue
show nerve endings (green)
aligning with muscle
receptor sites (red).

ALS victim uses an
implanted chip to communicate through a computer.

Different chemicals stimulate or inhibit different parts of the brain, thus providing treatments for various conditions. Special glasses permit a researcher at Eli Lilly & Company (left) to "see" the structure of the drug Prozac.

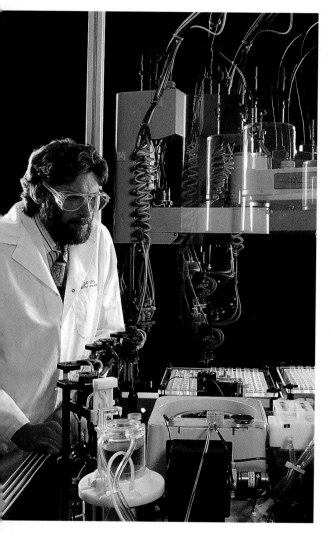

MOVING ARM of a robotic analyzer in a Lilly lab processes cultured animal cells that have been bioengineered to contain receptors for a specific neurotransmitter. By comparing how certain receptors interact with various chemicals, researchers develop new drugs to treat nervous-system disorders.

subtraction, this technique shows that the brain processing involved in reading and speaking is not distributed evenly throughout the organ, nor is it localized in one center of the brain.

"The functional areas involved in performing a particular task are distributed in several locations in the brain," says Steve Peterson, a colleague of Raichle's. "Each area makes a specific contribution to the performance of the task on the basis of its inputs, outputs, and particular information-processing abilities."

Within the decade, PET will be succeeded by functional magnetic resonance imaging (MRI) scans. MRI employs magnetic fields, radio waves, and computer reconstructions. Thus it is safer than PET, since no radiation is involved.

Functional imaging promises to provide a window on the normal brain as that brain's owner thinks, reads, or experiences emotions. But in order to achieve this goal of "mind-brain reading," functional imaging techniques must be combined with methods for measuring electrical events within the brain in real time.

While changes in oxygen and glucose use and in blood flow occur in time frames ranging from hundreds of milliseconds to several seconds, the brain's electrical signals travel from one neuron to another in tens of milliseconds. Computers linked to measurements of the brain's electrical activity are needed to manage information in these ultrabrief time frames.

Event-related potential recording (ERP) uses a computer to analyze voltage

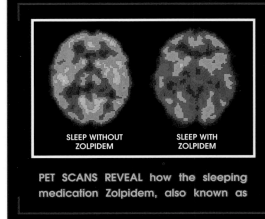

SLEEP WITHOUT ZOLPIDEM      SLEEP WITH ZOLPIDEM

**PET SCANS REVEAL** how the sleeping medication Zolpidem, also known as Ambien, affects different parts of the brain. Without medication, this patient suffers from chronic insomnia and sleep disturbances. Daily oral administration of Zolpidem helps restore normal sleep. Side effects of the drug, which is considered a short-term treatment for sleeping disorders, may include lightheadedness, nausea, and loss of memory. Severe insomnia that resists any treatment can signal an unrecognized psychiatric or physical disorder.

and brain-wave activity recorded from different points on the scalp. Using scalp electrodes, brain electrical activity is measured in response to stimulations as simple as a flash of light. ERPs reflect the patterns of neural activity that underlie mental processes like listening or looking.

While ERPs "capture" the temporal patterns of brain activity with precision, the method lacks the anatomical precision of PET and MRI technology.

"A collaboration is in the making," says Raichle. "PET and MRI working together will define the anatomy, and electrical and magnetic recording techniques will tell us about the time course of events."

**A**re there limits to how much can be learned about the brain over the coming decades? Technology will dictate how fast and how far neuroscience advances. But in addition to technology, new knowledge may depend as well on the answer to a paradoxical question: What are the limits imposed when, as in the neuroscientific enterprise, the subject of investigation and the investigator are one and the same, and the brain seeks to understand itself?

**HOW MUCH IS TOO MUCH?** An Eli Lilly researcher tracks the activity of a neuron with an oscilloscope and chart recorder. When stimulated, the cell generates the neurotransmitter dopamine. Too little dopamine is a symptom of Parkinson's. Too much can deeply affect mood and lead to psychoses such as schizophrenia. Most antipsychotic drugs work by blocking the brain's dopamine production.

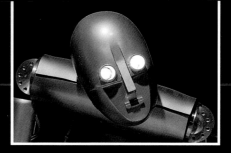

*Researcher holds the upended robot "Genghis."*

**S**omewhere between infancy and adolescence, the twin pursuits of robotics and artificial intelligence have made great strides—but they have not begun to approximate the complex circuitry that exists within the skull of a newborn human. Computer games test reflexes more than thoughts; chess grandmasters consistently checkmate the best computerized rivals.

Yet robots and artificial "brains" proliferate as human inventors—aided by ever more powerful

*"Gnat" robot*

microchips—increasingly blur the line between fiction and fact. Researchers at the Massachusetts Institute of Technology's Artificial Intelligence Laboratory model their designs after insects, reasoning that while they possess few neurons and limited mental ability, their shapes and simple behaviors have been time-tested for success. Hence six-legged contraptions such as "Genghis" (left), which "sees" through infrared sensors that detect moving, warm bodies and "feels" with pressure-sensitive metal whiskers. A microprocessor "brain" orders it to pursue moving things; place an object in its way, and Genghis either steps over or walks around. Legged robots, more practical than wheeled versions, may one day walk on Mars. MIT researcher Anita Flynn envisions earthly applications for her tiny creation, known as a "gnat" robot. She suggests it might serve routine but difficult-access jobs such as

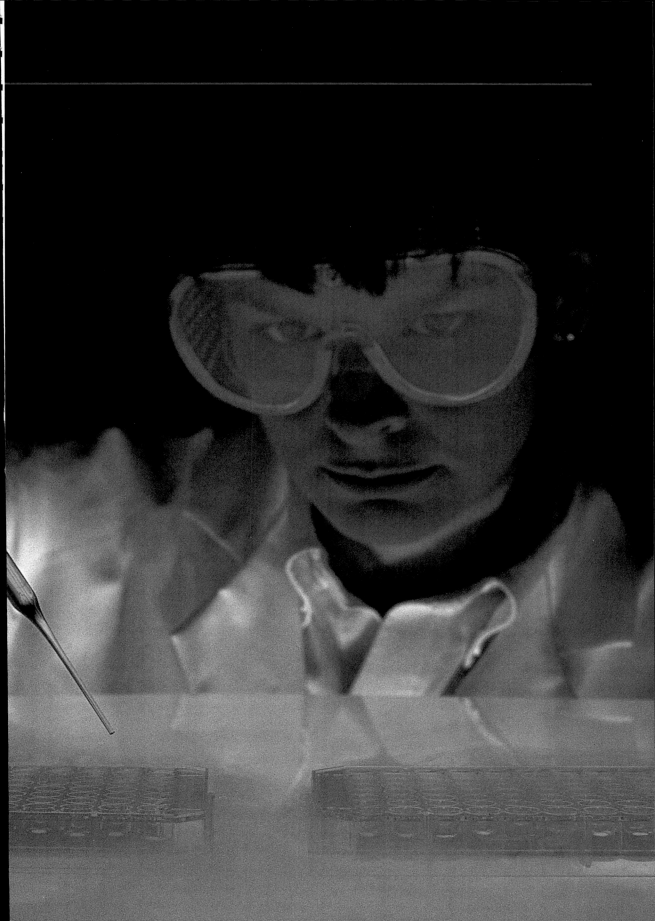

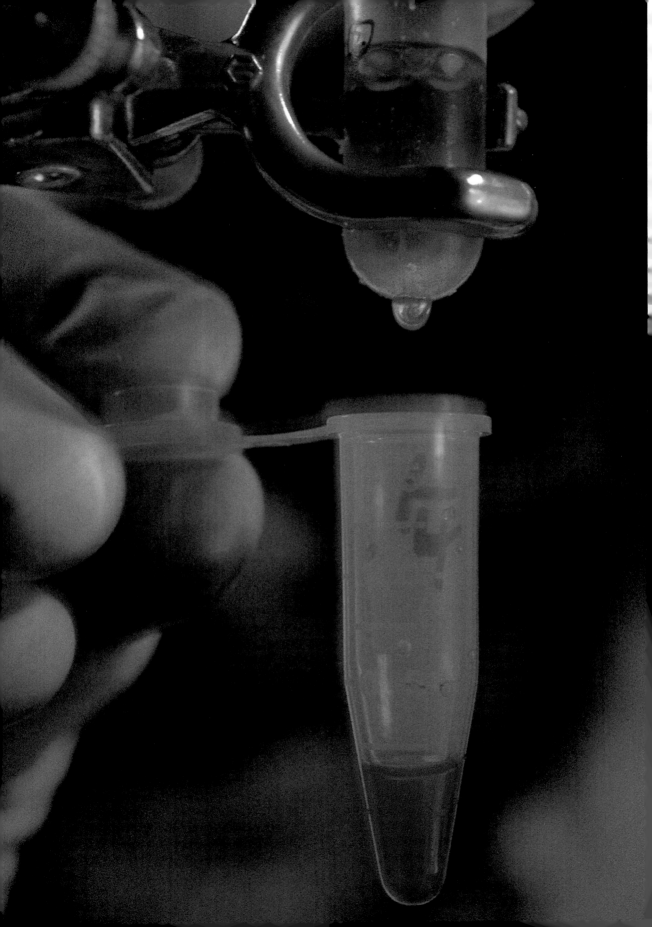

# NEW BIOLOGY

## 2

SEVERAL YEARS AGO, a young man in his late 20s and his wife walked into a medical building on North Wolfe Street in East Baltimore, where the Johns Hopkins School of Medicine sprawls upon a modest hill, and proceeded to the waiting room of a neuropsychiatry clinic. Like many young couples, the Gatewoods were anxious to start a family, but there was a little business to take care of first: a genetic test. John Gatewood's mother had died of Huntington's disease, a devastating neurological disorder that erodes vital and intelligent adults into involuntarily twitching, slurring, diminished shadows

The molecule of heredity found in the cells of living things, deoxyribonucleic acid—DNA—stores all the information necessary for life. A technician (right) examines a globule of DNA extracted from white blood cells.

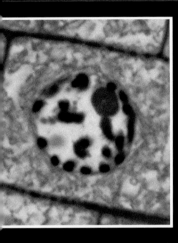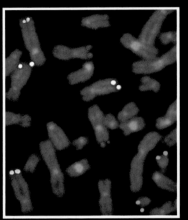

NUCLEUS OF AN ONION CELL (far left), magnified about 500 times, shows dark clusters of chromosomes containing DNA. Loose strands of DNA condense into these structures as a cell prepares to divide. On human chromosomes that are stained blue (left), fluorescent dyes tag certain genes along the DNA chain. The gene sequences are genetic messages that regulate cells and life.

AMAZINGLY ALIKE, a chimpanzee and a human share almost the same genes. Traits differentiating mankind arise from roughly one percent of the genes in the human genome. A computer-generated image of the DNA molecule (opposite) reproduces its structure—the double helix.

of their former selves. Although Mr. Gatewood knew that he had a 50 percent chance of inheriting the lethal gene, a series of genetic counseling sessions had made the ramifications of the disease clear to him. But in that mysterious way in which we sometimes think our fates could not possibly be ruled by the heads-or-tails statistics that govern inheritance, Mr. Gatewood had convinced himself that he probably had not inherited the gene.

What is it like to receive a genetic prediction? "Have you ever been in a car accident?" says Ann-Marie Codori, the

psychologist who counsels clients at the Baltimore center. "I remember the sensation as being stunned and shocked, sort of like you're not quite aware of what's going on around you. People are talking to you but you're not really hearing what they're saying. That's how people look when I tell them that they have the gene or they don't have the gene."

Many feel relieved of a burden, not only for themselves but also for their children. Some who test positive view the verdict as a gift; some who test negative feel guilty and are surprised that all the messy problems of life don't melt away. John Gatewood felt devastated. Not only did he learn with 95 percent certainty that he had inherited the gene; he also knew there was no treatment, no surgery, not even a drug in the pipeline to rescue him from this future. That knowledge plunged him into an emotional free fall. He became depressed and quarreled with his wife. He blamed her for pushing him to take the test. They no longer discussed having children. Instead they talked about divorce and ultimately broke up.

John Gatewood's life was turned upside down by what Columbia University genetics expert Nancy Wexler calls genetic knowing—the ability to peer into a human cell and read the runes of its nucleus as if it were a crystal ball. It may predict an agonizing death,

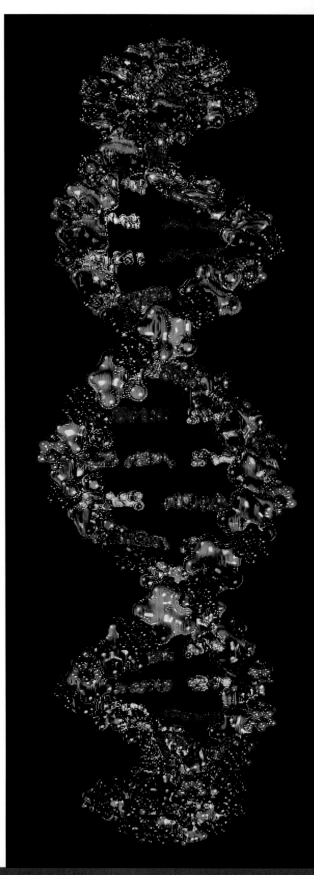

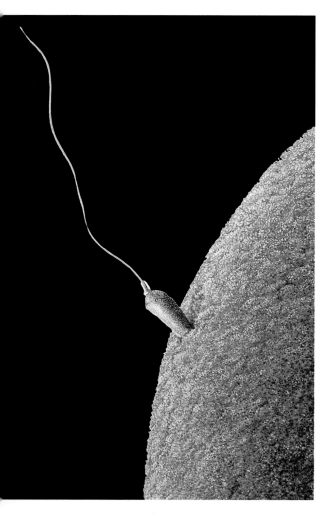

**AT CONCEPTION, a sperm penetrates the outer membrane of an ovum. Over a nine-month period, the embryo grows, and its cells specialize—forming the muscles, nerves, tissues, and organs of a human being.**

causes a rare, inherited form of colon cancer known as familial adenomatous polyposis (FAP). With this disease, hundreds or thousands of precancerous polyps begin to form in children as young as ten.

Gloria M. Petersen and her colleagues at Hopkins's School of Hygiene and Public Health began testing people in the mid-Atlantic region with a family history of FAP disease. One trip took Petersen to a young widow whose home had been wrecked by the rogue FAP gene as surely as by any hurricane or flood.

The woman's husband, father of her five children, had died in his early 30s of colon cancer, and a brother and sister of the man had the disease. The two eldest daughters, both teenagers, had already been diagnosed with the disease during regular examinations. When a genetic test for colon cancer became available, the mother decided to test the younger three children. The results were negative for two of the children. But the family's 11-year-old son had inherited the mutated gene, and physical exam nation confirmed that the cancer-forming process had begun. Within a week of learning the news, the boy's mother arranged for him to undergo surgery to remove his colon—a drastic form of preventive medicine, but one that doctors believe will give the child a better chance for a longer, more normal life.

Huntington's disease and FAP are relatively rare genetic disorders, but they are giving us a taste of both deliverances and dilemmas to come. Gene hunters are discovering a new gene a day; commercial

about which you can do nothing. Some of the details of John Gatewood's life, as well as his name, have been changed to protect his privacy, but his dilemma is all too real in an age when medicine's ability to predict inherited illness often outstrips its ability to heal our molecular wounds.

But you need travel no farther than the other side of North Wolfe Street to get a different glimpse of the genetic future. In 1991 researchers at Hopkins and at the University of Utah identified a gene that

A child inherits a full genetic endowment at conception. Unlike most cells, sperms and eggs carry a random mix of genes. The assortment is never the same, so, except for identical twins, each individual is genetically unique.

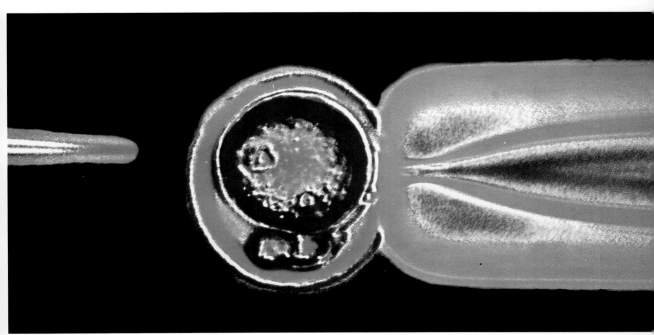

companies are rushing to the market-place with genetic-testing kits within months of these discoveries; and the more common disorders that touch us all, like heart disease and cancer and diabetes, are the focus of intense genetic explorations. With the predictive power of molecular medicine, we will decide whether to have children or not, what career to pursue, what foods we will eat, what habits we might change, how we plan to march into the future, and with whom. In a remark as appropriate to the era as to her personal test result, one woman told Ann-Marie Codori, "Life will never be the same." Neither will medicine nor biology.

Each of us has a key for traveling through one of those waiting-room doors,

GENETIC ENGINEERING: A slender pipette carrying DNA is about to insert a foreign gene into a mouse egg held by suction against a larger pipette. Test-tube twins (below) from eggs produced at the same time were born 18 months apart. One embryo was preserved and later implanted in the mother's uterus.

# HUMAN GENOME PROJECT

By the year 2005, this project aims to decipher the entire set of chemical codes that carry the blueprint for a human being. Scientists gain knowledge by knocking out genes in mice to observe the effects.

and it comes in the shape of the double helix. The idea that deoxyribonucleic acid, known familiarly as DNA, might be the essential molecule of genetic information was suggested back in 1953 by James Dewey Watson and Francis H. C. Crick, a discovery that raised the curtain on what is without question the most productive era in the history of biological research.

Pause for a moment and examine one of your hands. Each freckle represents a colony of roughly several thousand pigmented cells. Each of those cells possesses a nucleus, and each nucleus holds 23 pairs of chromosomes containing such a supercompressed stuffing of DNA that, if unwound and pieced together as a single thread, it would measure six feet. DNA is not remarkable for its length, however, but rather for the amount and kind of information it conveys.

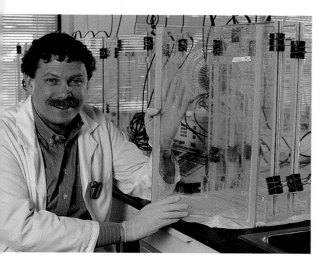

GENE MAPPER Eric S. Lander, Director of the Whitehead Institute Center for Genome Research in Cambridge, Massachusetts, holds glass plates sandwiching fragments of DNA in a gel matrix. Scientists cut DNA into useful segments, using naturally occurring proteins called restriction enzymes. These chemical scissors, which slice chromosomes at any desired point, are a basic tool in genetic engineering and in the search for genes that cause inherited diseases.

Just as a computer, using a minimal two-character "alphabet" of 0s and 1s, can steer a spacecraft through the solar system, DNA uses a minimal four-letter bio-chemical alphabet of A, T, C, and G to spell out every protein, every hormone, every molecule necessary for life. The letters stand for the chemical bases adenine, thymine, cytosine, and guanine, which join in strict pairs—A only to T, G only to C—like rungs between the two spiraling backbones of the double helix. Genes are, quite simply, discrete sequences of these letters; the exact sequence encodes information telling the cell how to make a particular protein according to strict rules of biological grammar. The three letters ACC, for example, always "spell" the amino acid tryptophan, one of 20 that join to form proteins, while ACT means "stop."

The complete and unabridged genetic text of humans—known as the genome—contains three billion letters of DNA. And perhaps the most remarkable

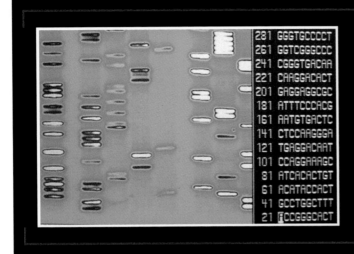

COMPUTER-GENERATED IMAGE portrays a sequence of nucleotides, or structural units of nucleic acids, along a fragment of DNA. Each bar represents one of the four base molecules, identified by enzyme reactions. A computer first sorts and then illustrates the results by placing the bar in the column for the chemicals A, C, T, or G. A single gene may consist of as many as 10,000 letters. The code itself doesn't reveal what the gene does.

and wondrous paradox of all is this: Any two human beings are 99.9 percent identical in terms of their DNA, and yet that 0.1 percent difference spells out all the physical uniqueness we see among humans.

The tremendous ferment in biology today, the rapid progress in locating genes, mapping them, finding out which proteins they make, testing those proteins as drugs, and creating three-dimensional atomic images of their structure—all this incredible burst of activity dates not so much from Watson and Crick's discovery as from the discovery of powerful new laboratory techniques known collectively as recombinant DNA. It was in 1973 that Stanley N. Cohen of Stanford University and Herbert W. Boyer of the University of California at San Francisco demonstrated that DNA could be cut up, recombined, and duplicated (continued on page 68)

AT THE WHITEHEAD INSTITUTE a researcher checks the frozen inventory in the human gene library. The small trays contain host yeast cells that are replicating, or cloning, human DNA.

FOLLOWING PAGES: A pointer with crosshairs guides a computer scanning genetic profiles.

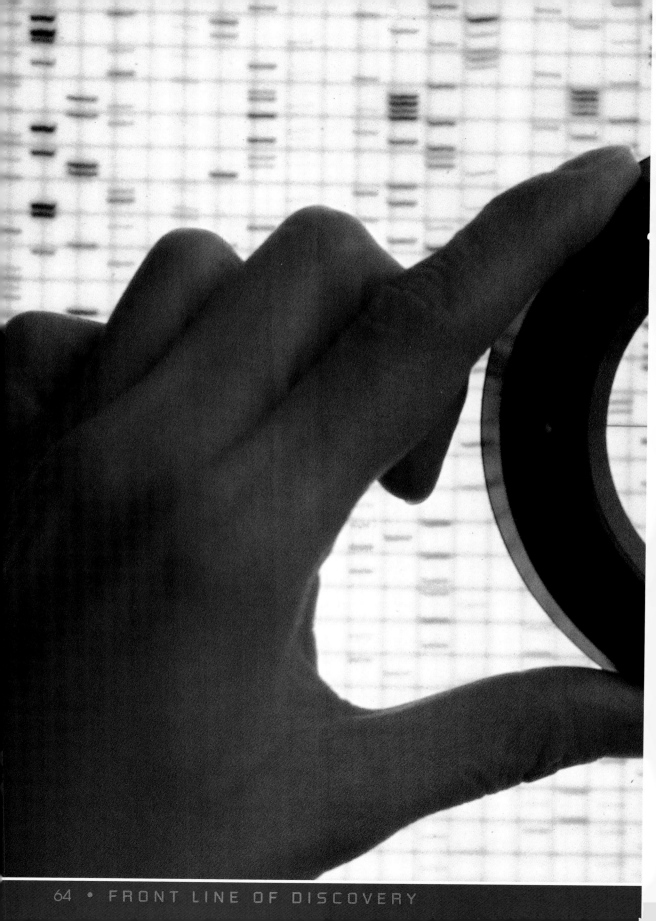

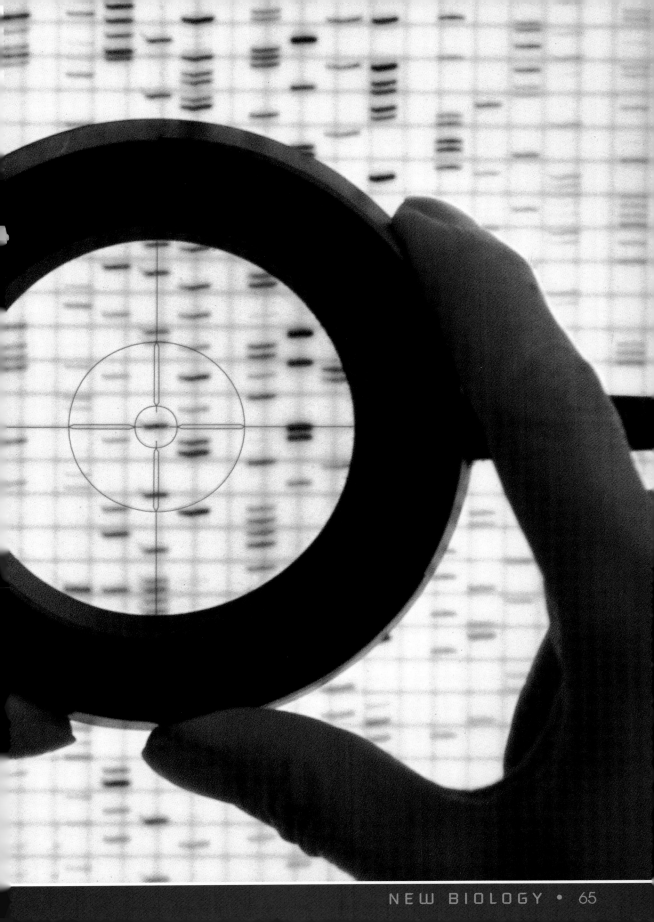

**B**reakthroughs in biotechnology promise a cornucopia of new plants—tastier, and more resistant to pests, diseases, and drought. For millennia, farmers have improved on nature by crossbreeding different varieties to produce desirable characteristics. Some 3,000 years ago, Chinese farmers cultivated a wild vine that yielded a black or brown soybean. An array of soybeans (opposite) selected from some 7,000 varieties illustrates the great diversity produced by conventional methods of breeding. Giving nature a hand by crossing strains of corn has improved harvests since the 1900s. Today tinkering

*Insect-resistant cotton*

with genes is faster and more dramatic as scientists transfer genes from one species to another and create carbon-copy plants in a process called cloning. To devise genetically identical pine trees, for example, scientists treat tissue with a solution of plant-growth hormones and induce it to grow until it sprouts. A geneticist at the Southern Forest Research Center at North Carolina State University (above left) examines tissue cultures of pine seeds taken from a fast-growing tree. The research may soon fulfill the goal of

*Cultures of pine seeds*

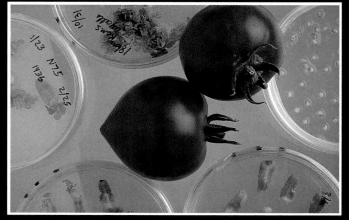

*Genetically altered Flavr Savr tomatoes (above); a selection of hybrid corn (lower right)*

growing designer trees with specific characteristics. Genetically altered Flavr Savr tomatoes spend more days ripening on the vine for more flavor. By isolating the gene that softens ripe fruit and plugging a copy of that gene into the plant backward, scientists have retarded rotting and created a tomato that does not have to be picked green. Experiments at Monsanto Company produce insect-resistant cotton. Inserting genes from certain strains of a soil bacterium into the plant's genes prompts it to produce a protein that thwarts pests like the caterpillar. Concern that altered plants might breed dangerous new strains of virus or disrupt

*Naturally grown varieties of soybeans*

natural ecosystems prompted federal safety guidelines, such as limiting access to test plots. As many thousands of experiments proved to be safe, early fears and regulations have relaxed.

**TO DETECT VIRAL DISEASE in chicks, researchers turned to a technique called recombinant DNA. The defective gene is cut from the virus and inserted into the DNA of a bacterium, which is cultured to make millions of copies of itself. The recombinant DNA acts as a probe, detecting carriers of the disease.**

in enormous quantities. With this sudden abundance of genetic material, other researchers showed how the letters of specific genes could be spelled out. Thus was the genetic text laid bare.

The culmination of two decades of such experimentation is the 3-billion-dollar Human Genome Project, launched in 1990 with the aim of identifying the exact location and biochemical sequence of every human gene in our six feet of DNA. Conventional wisdom suggests there are

100,000 genes scattered through this helical wilderness, but no one knows for sure; whatever the number, Harvard molecular biologist and Nobel Laureate Walter Gilbert has referred to this trove of information as the "grail of human genetics."

Perhaps a better way to think of it is as a trail. The trail of DNA, like some magic thread beckoning the biologically curious, leads in many tantalizing directions. It leads to new molecular medicines: Human insulin, growth hormone, and interferon, to name a few, are genetically engineered drugs that have been keeping people healthy and alive for more than a decade. The trail of DNA leads to the secrets of embryological development: Researchers like Sean Carroll at the University of Wisconsin have begun to

Making new breeds by manipulating genes could alter modern husbandry, as gene-splicing transplants DNA from one species to another. A bacteria-manufactured hormone, BST, increases milk production in a cow (left).

show how genes orchestrate the formation of limbs and appendages, such as the splendor of a butterfly's wing.

The trail of DNA can also lead to criminals: Forensic scientists at the FBI analyzed DNA from saliva cells on a discarded cigarette butt to help apprehend Sicilians accused of killing Italy's top anti-Mafia investigator in 1992. The trail of DNA can lead to sticky ethical and social dilemmas: Genetic discrimination is already a problem, according to Paul Billings of Stanford University, who has documented several hundred instances of people denied jobs or insurance benefits because of their genetic makeup. On a daily, routine, almost mundane basis, the trail of DNA leads to basic insights about biological processes: how molecules traffic within a cell, how genes turn on and off, which genes are linked to what disease, and how doctors might repair the damage. Most of all, the trail of DNA is leading biologists to an activity formerly the province only of geographers: mapmaking.

Today's genetic Magellans log their chromosomal landmarks on computer databases, and maps are distributed instantaneously through electronic mail.

This genetic atlas, and the biological world it depicts, is taking shape much more rapidly than people—including the biologists themselves—ever believed possible.

"Now to many people, that doesn't look so beautiful," Eric S. Lander was saying, motioning toward a map of all 20 chromosomes of a laboratory mouse. "But to me, it's a thing of beauty," Lander continued, "because what do you want out of a map? What you want is to be able to go anywhere, visit any neighborhood."

This mouse (continued on page 72)

**CLONED GOATS (upper right) may yield an advance in reproduction techniques. When the genetic material from prize animals is inserted into unfertilized eggs, surrogate mothers give birth to offspring with desirable characteristics. Fusing cells from a sheep and a goat produced a "geep" (lower right), with traits of both species.**

Cold storage conserves the genetic resources of prize animals and endangered species, preserving sperms, eggs, and embryos for the future. Vats of liquid nitrogen (right) store semen from champion bulls. Livestock scientists developed cryopreservation to selectively breed better milk cows or beef cows

Liquid nitrogen stores semen of prize bulls.

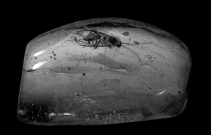

ENTOMBED IN AMBER, an insect retains pieces of DNA that can be extracted from its tissues. Scientists compare the DNA from such fossils to the DNA of their living descendants. The information can be used to measure rates of genetic mutation and to reconstruct the evolutionary history of plants and animals.

through artificial insemination. A number of zoos around the world are applying this technology to protecting endangered wildlife. Cylinders stored in liquid nitrogen hold living cells frozen at minus 385°F. Cells of such imperiled species as gaur, condor, and snow leopard are kept in suspended animation until needed for research and breeding. "If someone had done this with dinosaurs,

*Collecting DNA from tortoises*

we could possibly bring them back," says Betsy Dresser, Director of Research at the Cincinnati Zoo Center for Reproduction of Endangered Wildlife. "Frozen zoos can literally save a species from extinction." Dresser's center has achieved a first with its interspecies embryo transfers. In one procedure a domestic cat serving as a surrogate mother gave birth to an Indian desert cat. On another mission, veterinarians at the San Diego Zoo pin down a giant Galápagos tortoise (above) to draw blood for DNA samples. Scientists and scholars are hoping to increase the dwindling population of giant tortoises on Ecuador's Galápagos Islands. These heavyweights evolved separately on each of the islands and developed hereditary characteristics adapted to varying local conditions. Using blood samples, scientists produce DNA profiles that can distinguish these hereditary differences. By comparing the DNA profile of captive tortoises with those in the wild, scientists hope to repopulate the islands with the correct species.

*Frozen cells of endangered species*

BY COMPARING THE GENETIC patterns in families of gay males with those in the general population, a team of scientists discovered evidence linking genes to homosexuality. At the National Institutes of Health, biochemist Dean H. Hamer and a member of his research team scrutinize an autoradiograph for genetic markers shared by homosexuals as they search for the supposed gene or genes.

map, and its human counterpart, are taking shape as part of the Human Genome Project at the Massachusetts Institute of Technology's Whitehead Institute for Biomedical Research.

Lander, a tall and gregarious 37-year-old savant, directs the project. The immediate goal of this massive effort is to create two different maps of the same

DNA terrain, a genetic map and a physical map. A physical map can be thought of as a series of mileage markers that tells you where you are on a particular linear landscape, be it a chromosome or an interstate highway. The genetic map pinpoints the location of specific biological landmarks, ideally genes but also features known as markers, that may appear at

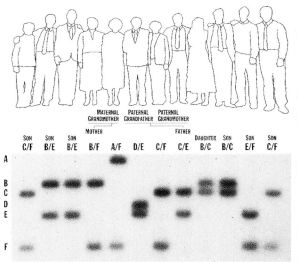

IN A PROFILE of the Guertler family of Salt Lake City, segments of DNA inherited by each member are shown as dark bands. The father's genotype is C and E, his wife's B and F. Their children inherited a letter from each parent in varying combinations. Geneticists use the carefully kept records of Mormons to confirm kinships revealed through DNA analysis.

## DNA AND HEREDITY

Almost foolproof clue to identity, a DNA profile depicts certain chemical sequences along a strand of DNA. These profiles establish hereditary ties and have reunited kidnapped children with their families (left).

irregular intervals along the highway. If we think of a chromosome, for example, as Interstate 40, which stretches from North Carolina to California, someone without a map might have to wander roughly 2,500 miles to find the exit for an important landmark like the Grand Canyon. But by breaking up I-40 state by state, county by county, even the ignorant sightseer can pinpoint the location of the Grand Canyon on a map of Arizona.

In much the same way, biologists start out as ignorant sightseers of DNA terrain. So they have taken to breaking up each chromosome into smaller, more manageable segments and homing in on microscopic landmarks. Indeed, the physical map is not so much a diagram tacked to a wall as a complete and ordered set of test tubes, each of which contains a standardized chunk of chromosome with regularly spaced "mileage markers."

The aim is to produce a guide to the location of all the genes on all the chromosomes, and the work is proceeding with astonishing rapidity. By 1994 the Whitehead's mouse group had produced a genetic map of that rodent totaling 4,006 markers—a goal they hadn't anticipated reaching until 1997. Why a mouse?

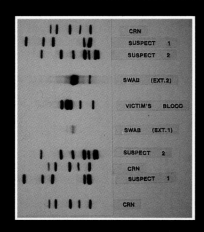

A SERIES OF DNA PROFILES makes a case against suspect 3, whose pattern of bands shows a match with the swab from evidence obtained at the scene of the crime. For DNA testing, scientists use only the DNA found between the genes where the sequences of the chemicals A, C, G, and T change from person to person. This DNA is chemically cut into fragments. Another chemical process splits the two intertwined strands of DNA apart. Radioactive probes made of single strands of synthetic DNA bond to matching chemical sequences, which for each person occur in different places along the DNA strand. X-ray film records the pattern, representing it in an autoradiograph that resembles a bar code.

"All mammals are extremely similar," Lander explains. "There's one basic body plan for the mammal. It was invented some one hundred million years ago, and there aren't significant differences among us. Our organs are in the same places, our brains work basically the same way, our immune systems work the same way. So the mouse is similar enough to us that it can serve as an experimental surrogate. If I discover a new disease gene in the human, the first thing I want to do is knock out that gene in the mouse to create a mouse model of the disease to study."

Enter the knockout mouse. Biologists have learned to knock a single gene of interest out of the embryo of a mouse to see what effect this selective genetic erasure has on the animal's development. A team headed by Susumu Tonegawa, a Nobel Prize-winning molecular biologist at MIT, recently reported, for example, that it had knocked out the gene for protein kinase C-gamma (PKC-gamma), an enzyme believed to be critical to strengthening the synaptic connections in the brain that are involved in learning and memory. What happened? When the knockout mice were tested, they dis-

played mild to moderate lapses in spatial learning while negotiating a water maze.

hile research on individual genes dominates current biology, many biologists believe the future's biggest challenge will be to tease apart multiple gene ensembles that collaborate to influence common human afflictions like heart disease, cancer, diabetes, and hypertension.

"What's most important is to retain a sense of humility and a sense of wonder about the whole thing," Lander says. "We barely know anything about the human genome. We're scratching the surface. Maybe for a few hundred genes we have some real clue how they act on their own. But there are 100,000 genes, and they collaborate with each other. There's a genetic symphony going on, and we shouldn't be pretending that suddenly we're maestros. At best we've met a few instruments and know what they sound like alone in practice. It's going to be a century—at best a century—before we understand how all these genes work together."

In this age of molecular genetics, perhaps the keenest area of interest right now is the most genetic of medicines: genes themselves.

Dr. R. Michael Blaese plopped his large frame down in his chair and officially pronounced it "one of those days." As an immunologist at the National Institutes of Health, he had just examined a four-month-old infant with a crippling and incurable genetic defect of the immune system called Wiskott-Aldrich syndrome. It was Blaese's unhappy duty to explain to the mother that there was little the NIH doctors could offer. "The hardest part of my job," Blaese said, "is writing letters to people saying there is nothing I can do."

Nonetheless, Blaese has considerably more to offer than back in 1990,

POLICE AT THE SCENE of a shoot-out (above) look for clues. A technician (below) lifts samples from a bloodied shirt. DNA profiling provides a new weapon in the war against crime.

FOLLOWING PAGES: Cells multiply in flasks containing a culture medium.

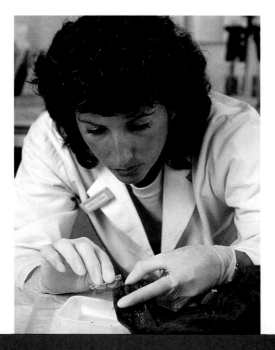

when he, with NIH colleagues W. French Anderson and Kenneth W. Culver, pioneered the world's first attempt at replacing a damaged gene, a procedure known as gene therapy. As more disease-related genes were discovered, scientists pondered the feasibility of replacing a damaged or missing gene. The NIH researchers focused on a serious immune disorder known as ADA deficiency; children with the disorder lacked the gene for the enzyme, adenosine deaminase. Without it, toxins build up and poison blood cells. "These children," Blaese says, "basically suicide their immune systems." They are known as bubble children, forced to live in (continued on page 78)

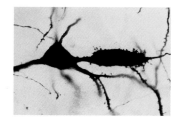

## GENETIC KNOWING

As scientists locate defective genes that cause inherited diseases, tests are developed to determine who is at risk. Such screening raises ethical questions about who has the right to know the stories foretold by DNA.

antiseptic isolation in the manner of David, the "bubble boy," who died in 1984.

David did not live long enough for gene therapy, but the approach to ADA deficiency was inspired by cases like his. The NIH team took advantage of a tiny microbe known as a retrovirus, which has the ability not only to infect human cells, but also to insert its volume of genetic information into the cell's own library of DNA. Using this virus as a kind of smuggler's bible, the scientists hollowed out the part of the virus's genetic text that let it replicate and inserted the gene for ADA, the missing enzyme. Then they harvested white blood cells from a young girl with ADA deficiency and allowed their engineered viruses to infect the cells. The neutered viruses ferried the missing gene into each cell they infected, and these cells—expanded in number to about one billion—could be reinfused into the patient, much as in a blood transfusion.

JEWS OF EAST EUROPEAN ANCESTRY, such as this Hasidic family, are at greater risk for carrying the Tay-Sachs gene, which distorts brain cells (top). Since the gene is recessive, it takes a double dose to produce the fatal disease. The defect may not reveal its presence for generations. Blood tests identify carriers with 100 percent accuracy. A new technique can now diagnose eight-cell, test-tube embryos before implantation.

The effort culminated on September 14, 1990. Blaese remembers the day well—"a time of great excitement, anticipation, and anxiety, I must admit. It was a big step. There was a lot of controversy about it. People said that it was crazy, it would never work, that we were cowboys." It was also a time of courage. The patient was a sickly four-year-old girl named Ashanthi DeSilva. "At the time when we first saw her," Blaese says, "she was experiencing repeated infections and was treated with antibiotics frequently. Her immune system was functioning very poorly." As Culver injected the re-engineered cells back into her body, Blaese and Anderson stood around worrying, hoping, waiting.

Nearly four years later, Ashanthi has regained a normally functioning immune system, with all the benefits that come with it. "Within a year of starting, her parents felt good enough about her

progress," Blaese says, "that they enrolled her in public kindergarten. Now she's in second grade and is doing wonderfully well. She has missed, I think, only one day of school in two years because of a cold."

Because of initial successes, gene therapy has become one of the hottest areas of medical research in the 1990s, with nearly 70 separate experimental approaches either under way or approved by the beginning of 1994.

Perhaps the severest test of gene therapy involved three newborn children on the West Coast. Prenatal tests showed all three fetuses had inherited the ADA deficiency, but babies offer a unique opportunity for a permanent genetic cure. Doctors know that the umbilical cord blood is rich in stem cells—the grandmother of cells from which all blood cells derive, including the immune cells damaged by ADA deficiency. For a brief period following birth, the infant's own stem cells circulate in the body before migrating permanently into the bone marrow. In May of 1993, acting quickly within this

A weapon in the war against genetic defects, recombinant DNA splices human genes into the simple DNA of bacteria, turning the bacteria into factories manufacturing human proteins, including hormones such as insulin.

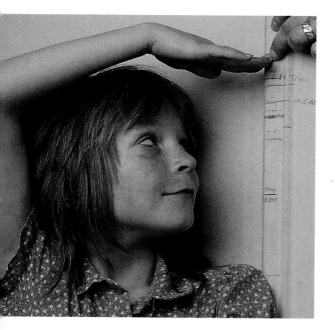

narrow window of opportunity, Blaese flew to Los Angeles with the retroviruses containing the ADA gene. There, doctors used it to infect blood cells harvested from the umbilical cord blood of all three infants; these cells, with the ADA gene now tucked inside, were returned to the infants. If the genes made it into any of the stem cells—and preliminary evidence suggests that they did—all three may enjoy normal lives.

"There's an enormous amount of anticipation about this technology," Blaese admits, his voice heavy with expectation. "But we tend to get excited about new things before they're proven. My crystal ball is a little cloudy as to what's really going to happen."

One of the most difficult tasks before us, scientifically and socially, is to resist the temptation to see all disease processes and their treatment as genetic, especially since much current gene research has in fact focused attention on environmental factors as well. Yet the sheer amount of information about genes will soon touch every household. "This new capacity," Robert Pollack writes in his recent book, *Signs of Life*, "has come upon us very quickly: it is as if we had just deciphered a few words in a new language and begun rewriting ancient texts before understanding their full meaning." We will use these texts, perhaps while they are still incompletely understood, to make some of our most profound life choices: whom we marry, whether to bear children, and how we will lead our day-to-day lives. How soon before this future is upon us? In one community, it is already here.

Indeed, in a 17th-floor office of Mount Sinai School of Medicine in New York, it began to take shape a decade ago. A distraught father named Joseph Ekstein was informed by geneticist Robert Desnick that his newborn had been diagnosed with Tay-Sachs disease, a severe and fatal neurological disorder that causes retardation and usually kills by age five. Ashkenazi Jews have a high incidence of the inherited disorder, and that was reflected in Ekstein's family. Three previous children had inherited the disease and had already died. "Is there anything we can do?" Mr. Ekstein pleaded.

There have been two answers to that question. First, with Desnick's help,

Ekstein established a premarital genetic testing program called "Dor Yeshorim" for teenage members of the Orthodox Jewish community. Participants who agree to be tested for Tay-Sachs are given a six-digit identification number and urged to call a phone number to check test results early in a relationship. If the couple is genetically incompatible, according to program director Howard (continued on page 84)

**A TEN-YEAR-OLD (opposite) born with a growth-hormone deficiency grew five inches in a year of treatment with the man-made human growth hormone, first engineered at Genentech, Inc. (above), in San Francisco. X-ray crystallography (below) shows the structure of the hormone and its receptors.**

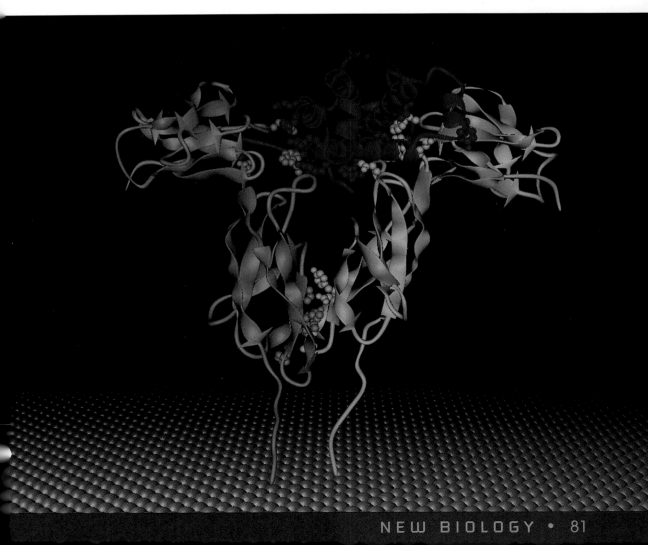

# IMMUNE SYSTEM

*Sneezing signals the invasion of a rhinovirus.*

*Magnified image of dust balls*

**G**uarding the body against disease, the immune system consists of an army of white blood cells, about a trillion strong, that are designed to search out and destroy invading microorganisms, foreign substances, and cancer cells. Sometimes the defense forces mount an attack against harmless invaders such as the particles in a dust ball. The response may cause cells to release infection-fighting histamine, which produces a runny nose, watery eyes, and perhaps a sneeze (left). At the front line of defense are white blood cells that circulate through the body, patrolling for invaders. White blood cells include macrophages, as well as T and B cells. A macrophage, or "big eater," (opposite above) reaches out to consume bacteria. If necessary, macrophages emit a chemical alert to helper T cells, which direct the immune response by recruiting killer T cells and steering them to the trouble spot. Killer T cells zero in on targets identified by their antigens and destroy them on contact. Helper T cells also activate B cells, which emit potent proteins called antibodies that

**GERM GALLERY: Electron micrographs portray the invisible world of viruses and bacteria, magnifying them hundreds of times. Tinted images show these unseen enemies that besiege us every day.**

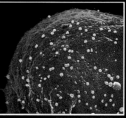

RESPIRATORY DISEASE

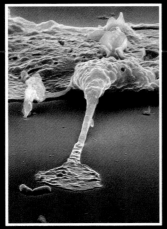

*Macrophage attacking bacteria*

neutralize the enemy and destroy infected cells. Suppressor T cells call off the attack after the infection subsides. Invaders also include fungi and protozoa, such as malaria parasites (below), which have multiplied in a culture of red blood cells.

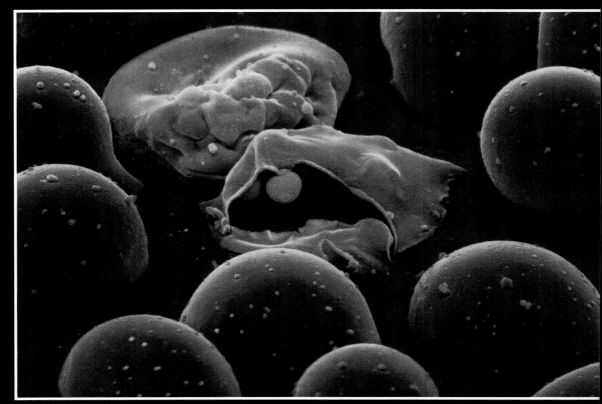

*Malaria parasites surrounded by red blood cells*

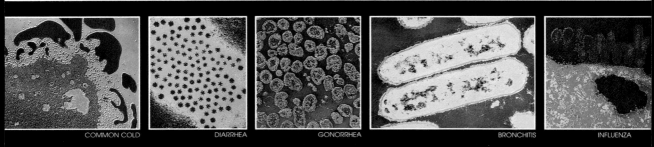

COMMON COLD    DIARRHEA    GONORRHEA    BRONCHITIS    INFLUENZA

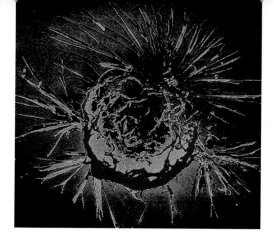

SINISTER MAVERICK, a breast-cancer cell (above) divides gradually at first, then grows wildly at the expense of surrounding normal tissue. By screening families with a history of premenopausal breast cancer, Mary-Claire King (opposite, at right) linked the problem to a region of chromosome 17. In September 1994, after a four-year search, a team of 45 researchers led by University of Utah's Mark H. Skolnick identified the gene that accounts for about 3 percent of breast cancers.

Katzenstein, "you haven't wasted that much time and emotion." In 1983, when the program began, only 45 people participated; now, according to Katzenstein, a total of 50,000 young people in the United States, Canada, Europe, and Israel have been tested. "We're up to at least 80 incompatible couples," Katzenstein says, "and our feeling is that the vast majority don't go through with the marriage."

The second answer to Ekstein's plea has grown into an ambitious research project undertaken as part of the Human Genome Project. Physicians at Mount Sinai are currently screening married couples from New York's large Ashkenazi community for up to three genes at once—those causing Tay-Sachs, cystic fibrosis, and Gaucher's disease—to study how people respond to genetic testing.

The readiness of this population to be tested may be paving the way for the

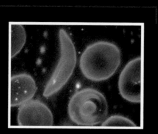

BLOOD CELLS in a crescentlike shape mark sickle cell anemia. A flaw in the hemoglobin gene produces an abnormal protein. Under stress, it comes out of solution, forcing flexible blood cells into sickle shapes that clog blood vessels. The defect in the gene that causes sickle cell anemia has been known for 25 years; still, no effective therapy or cure exists.

final frontier of molecular medicine, which according to Desnick and other scientists is the coming era of "predictive medicine": that is, screening individuals for the disease and susceptibility genes they carry and offering the opportunity to avoid, prevent, or treat genetic disorders. The Mount Sinai team foresees robots to analyze DNA quickly, and they hope to develop interactive computers to allow individuals to learn about genetic diseases and decide which tests they want to take.

"This will change the practice of medicine," says Desnick, who envisions the day when a person will provide a DNA sample and receive a printout of lifetime genetic susceptibilities for dozens of diseases. "It's likely we'll understand the functions of the human body at such a fine molecular level that we'll be able to devise treatments and cures for many diseases and conditions. You'll get your genome printout and a prescription for

## CANCER

**Failures in the immune system and DNA allow a renegade cell to become a deadly cancer. A breakdown in mechanisms that regulate cell division permits a mutant cell to multiply out of control and become malignant.**

improving or maintaining your health. Each morning, perhaps you'll take a health cocktail, tailored to modify your susceptibility to disease."

As we ponder this future of computer printouts and molecular medicine, how much do we truly want to know? It is worth listening to the voices of people who, in a genetic sense, *already* know:

"I have always lived one day at a time, and I think this has helped me to get through some of the days when things were hard for me," said one who tested as high risk for Huntington's disease. "In some ways I feel as though this is a gift."

"I wish I could be an ordinary person who wasn't plagued with these concerns for myself and family," said another. "Why did it have to be me?"

"Knowing will be better than not knowing," said a third.

The new biology brings with it a complicated, tricky, and surprising set of emotions. If the experts are correct, that future will be coming to a physician in your neighborhood soon.

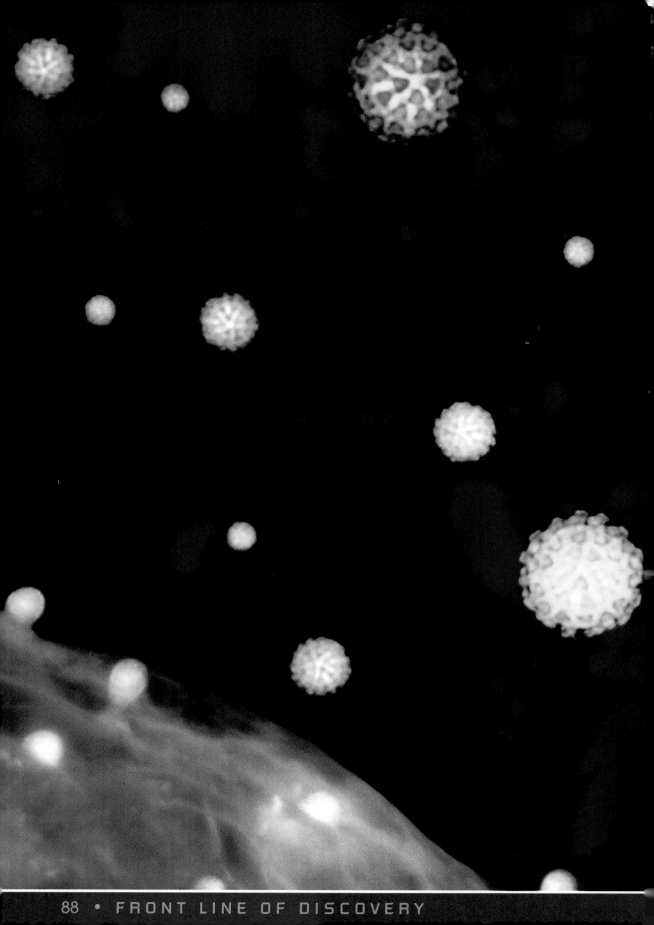

# NEW MATERIALS

## by Robert Friedel

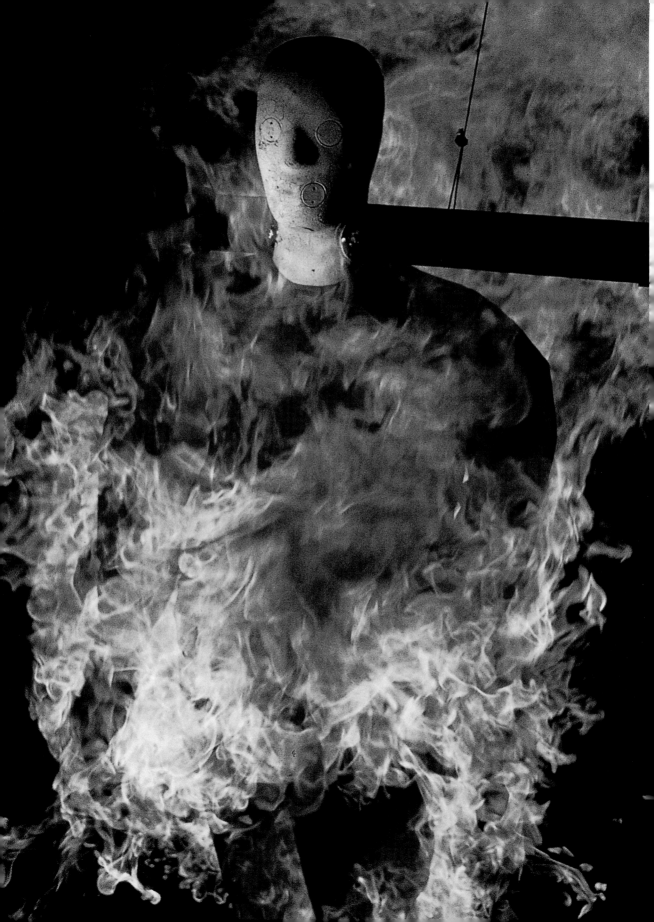

# NEW MATERIALS

3

ONE OF THE FIRST THINGS we discover
as children is that different things are
made of different materials. And we
learn that materials behave in different
ways. A wooden block feels harder
and heavier than a plastic one.
A paper plate doesn't break when
dropped; a china one does. Hit a
metal pan with a spoon and it makes
a clanging sound; pound on a card-
board box and it just goes "thump."
From these simple observations spring
a host of assumptions that we use as we
go through life. We think we know how
a material behaves, and that certain
substances are appropriate for some

Flame tests a flight suit of Kevlar, a tough, lightweight polymer.
PRECEDING PAGES: Optical fibers, sinew of the electronic superhighway

uses and not for others. These assumptions are so deeply rooted in our daily lives that even children rarely ask why, say, a piece of iron is hard and heavy while a lump of rubber is soft and springy. Most of us learn to conform to the behavior of the stuff around us and not to question it.

Materials scientists are different. Modern-day alchemists, they *do* ask why a material behaves the way it does. Then, when they start to get answers, they begin asking, "Why not?" as they make new materials that behave in ways different from anything under the sun. These advanced materials unleash a vast array of new technologies—from faster and safer airplanes to heart valves that can extend life for decades.

DIAGRAM of an atom shows electrons orbiting a nucleus. Later studies found the array to be much more complex. The nucleus is made up of even tinier particles—protons and neutrons. These particles are thought to consist of still smaller building blocks called quarks.

Sometimes the results are startling—as in a metal alloy so moldable it can be blown into a bubble, or a ceramic that goes into the heart of an automobile engine without melting or breaking, or a fabric that stops bullets, or a glass that is so clear you can look through a piece a hundred miles thick and not even know it is there. Just as often, the results are less flashy but equally important to economic well-being, personal safety, or the health of the planet we all share.

A recent development shows how important this can be. Take a moment to look at the book you are reading. It was written, edited, and published with the help of modern digital computers. Whether large machines or laptops, they are dependent upon microprocessors—computer chips that are the heart of modern electronics. Chips are made from crystals of silicon and small amounts of other elements. Silicon, an excellent material for controlling electrons, is a semiconductor.

These chips spawned a revolution in

**Modern research opened a window into the hidden world of atoms, the basic units of matter. With the new knowledge and new tools, scientists shape the building blocks into the materials and technologies of tomorrow.**

microelectronics because of their small size and speed. Without the ability to make and modify silicon crystals—a miraculous achievement of chemists and materials scientists about 40 years ago—the electronic marvels we take for granted would not exist.

So, excitement greets even the whiff of a major materials advance, such as superconductors or metal-ceramic composites or atomic-scale machines. Most of what follows is speculation, for it is difficult to predict how a new material will be used, or whether the hurdles of cost and technology will be overcome.

S uperconductivity, the flow of electricity without resistance, was discovered in 1911, but it was a phenomenon that could occur only at extremely low temperatures. In order to superconduct, metals needed to be immersed in liquid helium at 4 kelvin ( 4°C above absolute zero, or minus 460°F), making the process expensive and

impractical for everyday use.

In early 1986, IBM physicists Karl Alex Müller and Johannes Georg Bednorz in Zurich cooled a ceramic pellet of copper, barium, and rare earth lanthanum oxides and watched it lose electrical resistance at 30 K. This discovery set the scientific world off on a wonderful chase to find even better superconducting materials.

Twelve months later, a team headed by physicist Paul Chu of the University of Houston replaced the rare earth lanthanum with an yttrium compound. Chu's recipe lost resistance at 93 K—well above 77 K, the temperature of liquid nitrogen. This breakthrough was key: Nitrogen is abundant, easier to cool than helium, and a fraction of the cost. A superconductor

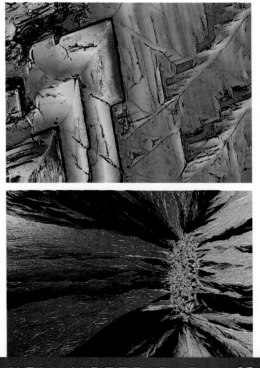

**SOLID MATTER in most nonliving things consists of crystals bonded in an orderly structure. Magnified crystals of Epsom salts (upper) reveal smooth, sharp-edged faces. A crystal (lower) appears in molten form. In a bubble chamber, particles too small to be seen (above left) create tracks in superheated liquid under pressure. Scientists use the chamber to study the behavior of subatomic particles.**

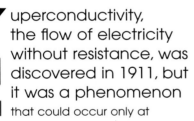

# SUPERCONDUCTORS

**They're in headlines, in supercharged research, in visions of trains speeding along on air. But the work lags as science pursues the right stuff—affordable and dependable, unresistant to the free flow of electricity.**

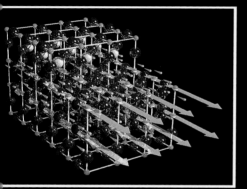

ELECTRONS STREAK through YBCO, the ceramic that loses resistance at 93 K. The model (left) presents a theorized atomic structure of YBCO: yttrium, barium, and copper in blues and gray, oxygen in red. Carbon (above, in crystal form) offers great promise as a superconductor in its new-found structure as a buckyball (upper right) or a buckytube (lower right).

that works above 77 K leaves the realm of the laboratory and promises to enter everyday life.

This promise not only jolted the scientific and engineering world; it also attracted unprecedented attention from the press, industry, and government. As a colleague expressed it, "Chu ran the four-minute mile in superconductivity."

"We seem to be poised on the verge of a new technological age," wrote Robert Hazen, a materials scientist and author at the Carnegie Institution of Washington, D. C. "In a similar way (to microchips) superconductors will gradually insinuate themselves into our lives."

"High-temperature superconductors" has become a catchphrase of the 1990s, springing loose money for research and fantasies about "flying" trains that travel 300 miles an hour using magnetic levitation, cheap electricity that can be stored

at will, and superefficient motors. However, the problems have been daunting. Ceramic materials of high-temperature superconductors are often brittle, unpredictable, and fickle. They tend to lose their superconductivity when anything but small currents are applied, and they lack the flexibility to be shaped into useful forms, such as wires or coils.

Innovative approaches are under way to overcome these obstacles. At the Massachusetts Institute of Technology, materials scientist John B. Vander Sande and his colleagues are making wires from easily shaped metallics, then oxidizing the wires into superconducting ceramics. At a nearby lab, MIT chemical engineer Jack Howard is exploring the potential of pure carbon materials, known as fullerenes—or buckyballs, named for the engineer Buckminster Fuller—to act as high-temperature superconductors. Long tubes of carbon atoms conceivably can be made cheaply and in large numbers; such buckytubes would be very strong, and under the right conditions could be made into superconducting wires.

Still, after more than half a decade of intensive research, no agreed-upon theory exists that explains the behavior of high-temperature superconductors, nor is there any consensus about how best to make such materials into everyday products. The high-speed trains, superefficient computers, and reliable power transmission systems promised by low-cost, high-temperature superconductors remain out of reach; but *(continued on page 100)*

**JAPAN'S EXPERIMENTAL MAGNETIC LEVITATION** train can hit five miles a minute, floating four inches above ground—but the levitating magnets require costly liquid helium for super-conductivity. A ceramic superconductor (left), wreathed in vapors of cheaper liquid nitrogen, suspends magnets above and below. Resistance in today's power lines dissipates billions of dollars worth of energy. Laboratories test many formulas (below) in a race for suitable high-temperature superconductors.

**FOLLOWING PAGES:** An electric arc between graphite electrodes creates pure carbon fullerenes, also called buckyballs, potentially the stuff of tomorrow's superconducting wires.

## METALS

Across the ages, technology advanced with metals. Today, new processes arrange metal grains to create tougher, easier-to-shape alloys, such as superplastic steel that can stretch 11 times its original length (left).

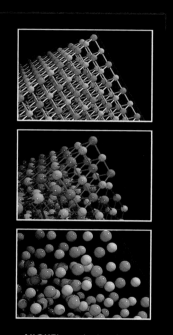

NICKEL and aluminum atoms align neatly in a single crystal (top) of nickel aluminide, a new metal of great strength. Combined crystals (center) may crack at the edges. Rapid solidification leaves atoms in glassy form (bottom), with altered properties.

when success is finally achieved, superconducting wires will be at the heart of such large-scale applications. The prospects make this one of the most exciting areas of materials science.

According to materials specialist Tom Forester, editor of *The Materials Revolution,* "There seems little doubt that superconductors are here to stay and that in time they will have a tremendous impact on technology and on society—and in ways that cannot be predicted."

The story of semiconductors and of superconductors drives home an important lesson about the power of modern materials. These substances, instead of being used for building things, are often used for controlling energy and information. Copper, which carried telegraph, telephone, and other communications for more than a hundred years, has

given way to glass, in the form of fiber-optic cables. The digital signals in these cables carry much more information than copper conductors of the same size or cost. Materials scientists are working now to make new kinds of glass that further reduce the costs of optical-fiber networks. They are making the glass so pure that light can travel in it for a hundred miles without any loss.

Some engineers, such as MIT's Lionel Kimerling, want to develop new forms of silicon that will carry light signals even more efficiently than the best glass. For Kimerling, silicon is the key to recent technological progress. "Electronic devices and circuits constructed from silicon have increased in performance and decreased in cost by a factor of one million during the past 35 years," he says. "Such an explosion of benefits is unparalleled in the history of any technology."

But further advances are needed, especially to support such grand ambitions as the information superhighway, now seen as crucial to the world's communications in the 21st century.

The manipulation of matter at the atomic level is one of the most extraordinary features of modern materials science. Only a decade or so ago, the proposition that we could see or measure matter at the atomic scale was considered impossible; the idea that

individual atoms could be picked up and moved seemed even more fantastic. But in the early 1980s, physicists at IBM's Zurich laboratory devised the scanning tunneling microscope (STM), which defied all presuppositions about what could be seen or done at the atomic level. The STM—and its cousin, the atomic force microscope, or AFM—allows us to "see" the atoms of a material, and even permits the positioning of individual atoms on the material's surface. In 1989, IBM demonstrated this capability by picking up and arranging 35 atoms to spell out the company's logo.

The instrument's ability to scan surfaces at the atomic scale has inspired efforts to design (continued on page 104)

**SMART WIRE NITINOL remembers its shape. Twist the nickel-and-titanium alloy into a spiral. Heat it (top left), dip it in ice water (top center), then reheat it (top right). Nitinol twists back into a spiral. Makers of springs and bra supports find uses for the "memory alloys." Nitrogen ions beam down on a racing-car crankshaft (right), hardening the metal without the potential distortion of high-heat treatments. Such advances project new chapters for the 5,500-year history of metals.**

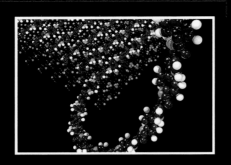

THE WAY THE POLYMERS lie—in long molecular chains, descending diagonally from the upper right of the computer model—gives strength to Kevlar fibers, five times the strength of steel by weight. In finished form Kevlar bears the brunt of harm's way, in bulletproof vests, flame-resistant suits, and boat hulls. The loop, enlarged, shows detail of the smaller molecules; each contains atoms of oxygen, hydrogen, and nitrogen (shown in red, white, and blue), around six carbon atoms tightly bound. Dupont, maker of Kevlar, combined the same elements in the 1930s to create nylon and neoprene, pioneers in the new world of synthetic polymers.

microscopic materials—built atom by atom. Called nanotechnology—from nano, a prefix meaning one billionth—many people now see this emerging field as a basic manufacturing technology of the 21st century. As IBM's chief scientist John Armstrong put it, "The ability to manipulate and to observe and measure things on that scale is tremendously exciting. We will have the ability to make electronic and mechanical devices atom by atom."

The astounding idea of creating microscopic mechanisms is not just fantasy. Researchers at Bell Laboratories, the University of California, Berkeley, and the University of Utah have made electric

motors the diameter of a human hair. Some minute devices and meters have already found their way into specialized applications, such as measuring the pressure in automobile engines or investigating how blood cells emerge from bone marrow. More advanced ideas include nanomotors that could be sent into the bloodstream to attack viruses or even nanorobots that could be programmed to replicate themselves.

Promoter of this new technology and its consequences, K. Eric Drexler claims that we are moving "toward an era of molecular manufacturing giving thorough and inexpensive control of the structure of matter." Some day, he says, molecular manufacturing will be able to make everything. According to Drexler, nanotechnology will completely transform information technology, biotechnology, and materials science, enabling us to build self-replicating engines of abundance, engines of healing, and engines of destruction. Not everyone agrees—some feel that nanotechnology is more in the realm of science fiction—and the debate over its real implications is a heated one.

**LIGHTEN UP, BRIGHTEN UP, with polymers:**
**Silvered polymer film on a metal membrane**
**makes a mirror lighter than glass and easier**
**to aim at the sun; the heliostat focuses rays**
**to produce energy.**

**FOLLOWING PAGES: Racers find polymer-**
**treated turf at Oklahoma's Remington Park**
**fast, soft, and never muddy.**

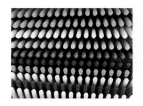

## POLYMERS

## POLYMERS

The name polymer, from Greek, means many parts—small molecules chained into a large one. They also have many uses in nature and in synthetics, including plastics, such as the many-colored combs at left.

Other implications of working at this tiny scale are not open to debate and are beginning to transform a wide range of materials, from metals to ceramics. Materials makers know that many of the important properties of a substance—its strength, elasticity, and durability—depend upon how the atoms or molecules of the material hold together. In most solids, such as metals or ceramics, these particles form crystals, and the crys-tals cluster together as tiny grains. The size and arrangement of the crystals and grains give the material its essential char-acter. When we change the size and arrangement, we make new materials with new properties.

The distinctive behaviors of different forms of steel, for example, turn out to be governed largely by how the grains of iron, carbon, and trace elements in the steel are *(continued on page 112)*

*Glass fiber in a needle's eye*

Labs a time exposure (opposite) makes visible a simplified demonstration of how varying pulses of light and dark stream bits of information through a fiber. A home computer with a telephone and modem can cruise the fiber-optic byways of Internet, to chat with fellow travelers in 60 nations, to tap into troves of knowledge in great libraries and universities, to ask questions or answer them. The potential, says computer scien-

tist Stephen Steinberg, is there for "a global brain-bank...a modern agora...a place where we can all be heard." Bell Labs' Rapport system (below) shows how members of a group from around the world can all be seen. With computer, phone, and tiny video camera beside the monitor, colleagues converge for a desktop conference. Fiber optics and more recent advances in digital transmission and signal compression have

*A multimedia conference*

**T**he information superhighway beckons. Not all of it just yet, but some fast lanes lie open by virtue of glass fibers so slender that they can thread a needle. Through one of these wisps fiber-optic transmission can whisk the words of 200 books in a second. Over a single pair, 10,000 calls can take place at once. At Bell

brought nearer the day when cable TV or phone companies can flood the superhighway with 500 channels. They will offer popular films, home shopping, and more. .

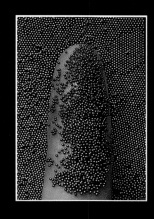

ONE OF THE EARTH'S MOST abundant elements, silicon (left) forms the main ingredient of glass, including optical fiber. New kinds of silicon that carry light signals more effectively than the best glass will be crucial to the electronic superhighway.

*Encoded in a light beam, data flow through optical fiber.*

# BODY PARTS

*Arm wrestling with a myoelectric arm*

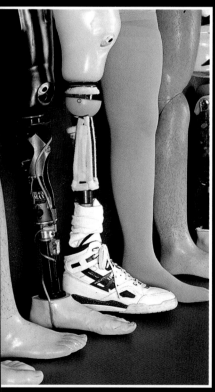

*A selection of prosthetic limbs*

springy plastic made Nebraskan Roger Charter (right) the first double amputee ever to run. When cowboy Bob Goodman (left), victim of a high-voltage cable, flexes arm muscles, skin sensors send signals to move his myoelectric arm and hand, developed at the University of Utah. He's now back in the saddle working his Oregon ranch. Growing apace

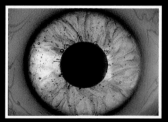

*Prosthetic eye*

**H**ow does it feel? With sensors on the ball and heel, according to designer John Sabolich (lower right), as he demonstrates in his pioneering prosthetics workshop. Pressure on the sensors, sending an electric tingle to the wearer's skin, offers a semblance of feeling. In place of heavy wooden legs, today's limbs promise user-friendlier form and function. Advanced metals, composites, and plastics all contribute. Knees and ankles of titanium, resin-composite shins, and feet of a new,

with the materials revolution, the human body shop expands its wares to replace worn, damaged, diseased, or defective parts—and to naturalize them. In clinical trials, natural bone grows smoothly into a "bioceramic" material. It could extend the life expectancy of hip replacements (upper right), which grow loose-

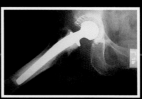

*X ray of a hip joint*

jointed over time. Seed cells from circumcised skin grow in a nutrient medium on dissolvable netting (right); doctors hope the dermal patch will aid patients with skin ulcers and reduce scarring and the long agony of skin grafts endured by burn victims.

*Lab-grown skin cells*

*New limbs enable a patient to run.*

*Designer Sabolich displays pressure sensor in a prosthetic foot.*

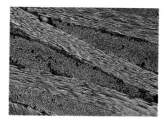

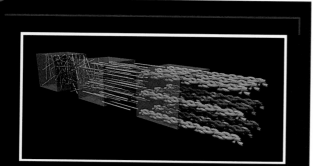

FIBERS—JACKSTRAWED, NEATLY STACKED, braided, or in a variety of other patterns—play the major role in making composites, pound for pound, the strongest of all materials. Fiberglass boats use short, chopped fibers. Parallel fibers in layered composites help lighten airframes. Rocket launchers may use interwoven fibers. Boeing's brand-new 777 jetliner took to the sky in mid-1994, lighter than previous planes by 2,600 pounds, thanks to structural composites.

or welding. In its most plastic form, Sherby's steel can be blown into bubbles. More important, it can be molded into intricate shapes, requiring less machining than normal materials.

Building materials atom by atom benefits nonmetals, too. Nanophase ceramics can be shaped more easily than ordinary ceramics and yet resist the shrinkage that usually plagues the finishing stages of ceramics molding. At the

arranged. "Nanophase" grains of metal—no more than ten millionths of a meter across, about a thousandth the size found in ordinary steel—can be pressed together to make superdense materials that are several times stronger than traditional metal. High-strength, low-alloy steels are widely used in structures where extra strength and reliability are called for—such as bridges and oil platforms. Other superfine-grain steels are even more surprising. Oleg Sherby, a metallurgist at Stanford University, has created a super-plastic steel that is so moldable it eliminates the need for traditional machining

**Materials matched and mixed for lightness, strength, or good wear, go into skis, racquets and golf clubs, autos and aircraft. Glass fibers (left), magnified 62 times, form blurry streaks in a polypropylene matrix.**

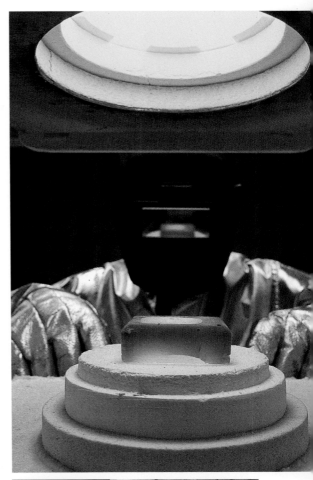

Argonne National Laboratory, outside Chicago, Richard Siegel is producing superdense nanophase ceramics that can be bent like plastics. These materials are slow and expensive to make now, but Siegel sees no reason why they can't be made in economical quantities to be used as computer chips and other electronics.

The practice of "thinking small" especially affects the composites—materials of remarkable strength and lightness. Ever since straw was added to brick in biblical times, people have understood that one material can reinforce another. In our everyday lives, we encounter composites in everything from bathtubs and fiberglass boats to tennis rackets and fishing poles.

Now, carefully engineered reinforcing elements are used to change the properties of brittle or weak materials like ceramics or concrete. A remarkable example is the metal-ceramic composite, known as the cermet. At the Lawrence Livermore Laboratory in California, Danny Halverson, one of the developers of a combination of aluminum and boron carbide, originally had in mind to make a

**FRESH FROM THE OVEN descends a cermet (upper)—composed of a ceramic and a metal. Created in a quest for lightweight armor, the recipe may be best suited for "wear parts," such as ball bearings. A scientist (lower) ignites a chip made of metal layers so thin that thousands together scarcely match the thickness of a human hair. Light-skinned and reusable, the revolutionary DC-X (left), in a scale prototype, blasts off.**

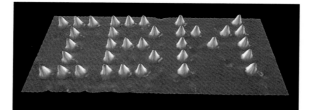

superior to some of the best tungsten steels—yet one-third the weight. Halverson points out that "probably the best applications are wear parts—simple geometric components that must resist high wear, such as pump seals or ball bearings."

Other cermet types of composites may be more economical, relying on cheaper ceramics and ingenious manufacturing techniques. At the Lanxide Corporation in Delaware, materials engineer Marc Newkirk works with aluminum alloys. He turns their natural tendency to form oxides from a nuisance—oxides are a form of corrosion—into a clever means for mixing ceramics with metals in an economical process. According to Newkirk, products result that combine "the toughness of steel with the lightness of aluminum." These new materials can successfully compete in both properties and price with traditional molded or machined metal parts, such as turbine blades and mining tools.

Lawrence Livermore scientist Troy W.

**A RESEARCHER USING** a scanning tunneling microscope (STM) "feels" gold atoms with a feedback device. The microscope lets him manipulate the gold at the atomic level. Using the STM to move xenon atoms, IBM crafted its logo (top), magnified 3.3 million times.

lightweight material for military armor. By combining a powder of boron carbide, one of the hardest materials known, with aluminum, and then heating it, Halverson created an incredibly tough material,

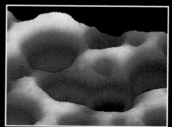

**NANOIMAGERY,** courtesy of the scanning tunneling microscope, presents false-color molecular landscapes. The STM scans with a superfine needle and maps the atomic contours by recording the rate of electron flow between tip and surface. The image produced shows topography of chemical elements and compounds revealing location and arrangement of individual atoms.

POLYMER ALLOY

## MOLECULAR LANDSCAPE

A new microscope, worthy of a Nobel Prize, opened the eyes of science to nanoscapes—the shapes of atoms. Researchers now can move atoms, build molecules, and envision nanomachines.

Barbee is leading a team developing another composite that takes advantage of our ability to alter matter at the atomic level—what he calls "atomic engineering." His "multilayers" are sheets of metals and metal combinations formed by layering astonishingly thin films—each only a few atoms thick—of alternating materials, until several thousand layers have been put together. These layers total perhaps a hundred microns in depth—a bit more than the diameter of a human hair.

By a technique known as sputtering, Barbee alternates layers of, say, copper and copper alloyed with zirconium, producing a sheet that has much more resilience than either of the original materials alone. A multilayer of copper and copper-nickel alloy, with as many as 5,500 layers, possesses a tensile strength more than ten times that of copper.

Multilayers—which have been made with as many as 40,000 layers—can combine two properties, such as hardness and toughness, that often are not found together. This makes them useful for everything from integrated circuits to high-speed engines. The multilayer technique also is used to make carefully designed metal-ceramic composites, in which the material starts as a ceramic layer, then gradually, layer by layer, results in a top layer that is metal. These composites can withstand heat like a ceramic and be flexible like a metal—potentially useful for making more efficient automobile engines. As Barbee says, they are truly "designer materials."

The ultimate designer of materials—Mother Nature—has been working at her designs for millions of years. According to George Haritos of the Air Force's Office of Scientific Research, she has gotten rid of her failures, leaving "the best she can offer." Everything from *(continued on page 118)*

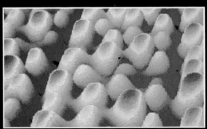
SILICON ATOMS

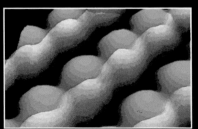
GALLIUM ARSENIDE

BENZENE MOLECULE

# MICRO-MECHANICS

*Micro-gear; actual size, a hundredth of an inch wide*

Consider the ant, in life, crawling across this page. Downsize the gear to scale. Now a speck, it scarcely serves to dot an "i," this cog in the micro-industrial revolution. A German lab made it—and also a 2,500-rpm turbine that fits within a hairsbreadth. From layers of silicon Berkeley scientists built the first rotating micro-motor in 1988. The microscope (top right) scans a wafer laden with 7,000 of them. To test one of the fragile machines, four fine probes (right) nudge it into place. Among many uses, silicon micro-sensors monitor fuel efficiency in car engines and measure the pressure on heart valves and in tires and scuba gear. From micro-machines to nanomachines, the field shrinks a thousandfold—from millionths to billionths of

*Researcher scopes micro-motors like one in projection.*

a meter—and expands to a vast, complex, and visionary world. K. Eric Drexler, dubbed "Mr.

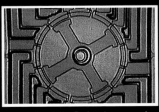

*World's first micro-motor*

Nanotechnology," co-designed the bearing at right for a nanomachine; molecular tools with chemical tips would position the 2,808 atoms. It has not yet been produced. Nanotechnology builds on advances in physics, biology, chemistry, and computers. Fascination builds on theoretical vistas; there loom nanomachines that one day may fashion a hand-held supercomputer, store every word printed since Gutenberg on a wallet card, travel through our bodies making microscopic repairs. As molecular tools beget better ones, Drexler foresees them building skyscrapers, even cities. Dream machines? Futurist hype? Some in the field think so. Still, five centuries ago Leonardo sketched flying machines. What could it be but a dream?

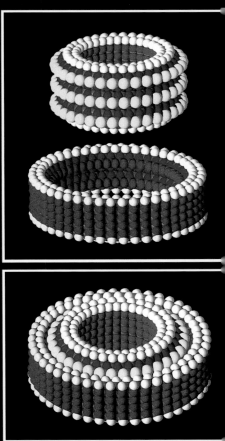

*Molecular bearing design; inner ring rotates.*

*Testing a micro-motor*

*Silicon micro-sensors*

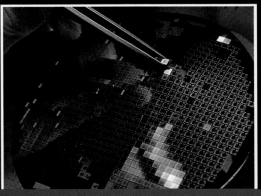

Horn, web (left), and shell—natural composites, polymers, and ceramics—inspire a new field of research. Biomimetics studies winners in the wear and tear of survival, seeking to adapt their strengths to new materials.

the silk of the spider's web and the outer covering of beetles to the tough shell of the abalone has come under the materials scientist's scrutiny. Few usable products have emerged from this research. Some scientists, such as Penn State University's Rustom Roy, have voiced skepticism about the promises made for them. Even so, the field of biomolecular materials is a hotbed of exciting investigations.

Perhaps the most widely pursued research involves the silk produced by certain spiders. Investigators are trying to learn what gives this material its strength, toughness, and resilience. The University of Washington's Christopher Viney has discovered that the protein of a spider's silk begins as a liquid crystal, such as that used for digital displays. At the U. S. Army's research laboratories in Massachusetts, David Kaplan and his colleagues have shown that when the spider secretes the liquid, the molecular arrangement of the protein is changed, making it insoluble, elastic, and stronger than steel.

The investigations to understand how these properties occur and how they can be copied involve chemical, mechanical, and genetic research. Ultimately, a material may result that is as strong as such artificial fibers as Kevlar and as tough as steel; a material that could be used for anything from bridge cables to tank armor.

"We are still in a fundamental stage of research," says Viney, "but the silks are teaching us about the molecular architecture needed for attaining strength and toughness in a material." And, as Viney and Kaplan both point out, a silklike material could be made without pressure, heat, dangerous chemicals, or harmful wastes—

WOOL HAS COMFORTED HUMANS for 12,000 years. Its scaly covering (seen from left in coarse and fine sheep's wool, alpaca and cashmere, with silk, linen, cotton, and polyester) sheds moisture while the core absorbs as much as a third of its weight in liquid. Scientists are trying to mimic this quality for use in synthetic fibers.

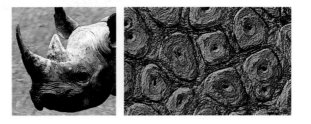

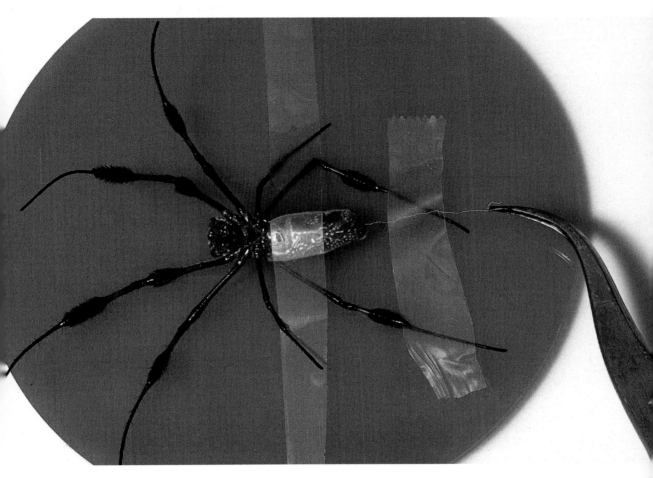

A SPIDER YIELDS SILK (above), a stretchable polymer stronger than steel or synthetic Kevlar. Scientists seek to synthesize it, perhaps to make sutures and ski suits. Natural composite of rhino horn (top right and left), similar in structure to the material used in Stealth aircraft, has remarkable healing qualities. It can self-mend tiny cracks that result from confrontations with other rhinos.

an environmentalist's, as well as an engineer's, dream.

The effort to copy nature's materials is called biomimetics. One line of investigation involves the abalone shell. Shells are made largely of inorganic minerals, essentially ceramics, that are held together by protein binders. Some shells, such as the abalone's, possess incredible strength for their delicate thinness—a full-grown man can step on a shell and not break it.

**SOLID LIGHTER THAN AIR, SEAgel floats on bubbles; only the air in its microscopic pores keeps it from floating away. In denser form it can function as packaging or insulation, a potential competitor to foam plastics. Water and bacteria swiftly decompose it, say its creators at Lawrence Livermore National Lab. The gel derives from kelp, the seaweed that grows in California's submarine gardens.**

It turns out that this strength is due to the microstructure—just like that of the engineered ceramic composites. But nature has found a way to produce the microstructure of the abalone shell without heat, high pressure, or the exotic components of man-made ceramics.

As Princeton University materials scientist Ilhan Aksay has commented, the ingredients of the shell are remarkably ordinary—calcium carbonate, or chalk—but the result is "as strong as the most advanced man-made ceramics." If the means for duplicating the design of the abalone shell could be applied to even tougher components, the result could be the strongest and toughest ceramic yet.

With nature as a model, scientists are learning how to improve the properties of materials, as well as to make substances

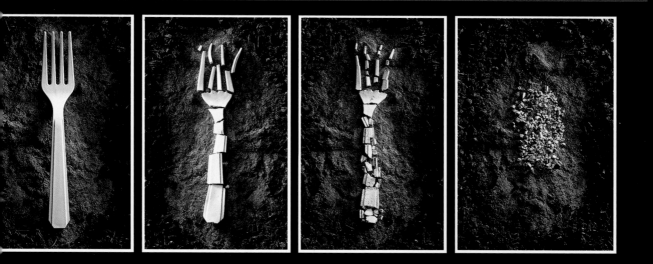

# CLIMATE

by Carole Douglis

# CLIMATE

# 4

THE TREASURE HOUSE is huge and, at 40° below zero Fahrenheit, may be entered only after donning the appropriate ritual robes. Feeling virtually upholstered in an army-issue polar snowsuit, I accompany Tad Pfeffer and Geoff Hargreaves inside. Fluorescent lights glint off silver rods, row after row, floor to ceiling. Treasure rests inside each of the aluminum cases. It consists of a six-inch-diameter cylinder of ice, called an ice core. The core has been drilled from the top of Greenland's ice sheet all the way down to bedrock— so there are three kilometers of it, sliced into one-and-a-half-meter sections.

Ocean currents circulate heat around the globe; red indicates warmer flows.
PRECEDING PAGES: Fragile bubble of our atmosphere evolved along with life itself.

CLOUDS' ROLE IN CLIMATE is not fully understood: Some warm the globe by trapping heat; some cool it by reflecting sunlight back into space. At any one time, clouds (above and right) sheathe half the globe's surface.

Preserved in the core are water and air from snowflakes that began falling 250,000 years ago. Engineers would be hard pressed to design a better record than an ice core, say those who study them here at the National Ice Core Laboratory in Lakewood, Colorado.

The ice holds whatever was blowing in the wind at a particular time—the precise composition of gases, dust, volcanic ash, pollen, and, more recently, pollution. Physical properties of the water will tell

years old. Along with microscopic marine fossil remains, the cores contain soil, dust, and volcanic ash blown off the African continent: The cooler and drier the climate, the more soil and dust in the wind, so the thicker the layer of sediment. Sure enough, cores dating from 2.8 million years ago show thick layers of dust on the seafloor, indicating cooler, drier, and windier conditions on the mainland.

hy might colder conditions be an impetus for such bursts of evolution and increase in brain power? Says Vrba: "Visualize the pressure put on organisms by natural selection for finding a solution to survival in cold weather. They may have to learn to do new, 'clever' things, like store food from one season to the next. Or migrate. Social interactions then become more complex, requiring lots of complicated communications—because the environment is so tough to beat."

One intriguing hypothesis that Vrba and others are researching: The fossil record shows that during cold periods, many mammals evolve larger bodies, spend longer in the womb, and, as they age, retain the relatively high brain-to-body proportions characteristic of mammalian young. Could this process, called juvenilization, help explain why human adults can retain youthful curiosity, inventiveness, and playfulness? Could it help

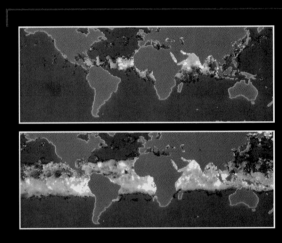

THE JUNE 1991 ERUPTION of Mount Pinatubo in the Philippines shot two cubic miles of fine ash into the atmosphere, along with 20 million tons of sulfur dioxide. In the wake of the eruption, satellite data (top) show increased sulfuric aerosols primarily in the Indian Ocean. Two months later (bottom) an "aerosol parasol" had encircled the globe. The sulfuric aerosols reflected sunlight back into space, temporarily lowering global surface temperatures by about one degree.

explain the unique creativity that has not only led to our artistic achievements but also helped *Homo sapiens* become super-adaptable and dominate the globe?

Right now these questions can only be answered with educated guesses. It is clear, however, that climate shifts left their stamp not only on our bodies but also on the development of civilizations.

In a computer-crowded lab, Jim White hands me *(continued on page 140)*

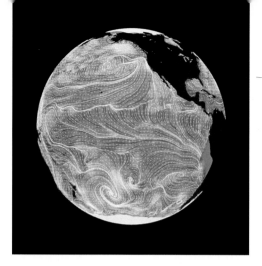

**FROM ABOVE THE EARTH, NASA satellites** monitor wind speed and track storms around the world. On a particular day, the above satellite map showed extreme wind velocity in the South Pacific and off the Alaskan coast.

a graph bristling with erratic spikes. "This is the first time we've been able to look in detail at a full glacial-interglacial cycle," he says. The graph shows average global temperatures over the last 150,000 years—more than the life span of *Homo sapiens*.

Most of the page resembles an EKG gone wild. White points to the part that looks not only warm but by far the calmest: "This section represents temperature ranges during the last 10,000 years," he says.

This 10,000-year interglacial also happens to be the relative blip of time during which civilization as we know it flowered. This may not be a coincidence: "Can you imagine trying to start farming for your food—and have the climate change drastically after a few years?" White asks. Archaeologists now think that agriculture sprang up early in the interglacial in various places independently. Perhaps people had tried planting crops many times before, but their efforts couldn't "take" until the climate settled down.

Climatologists have known that temperatures fluctuated dramatically during the last ice age. But the Greenland ice cores revealed some new information: Rather than taking place over centuries or millennia, it looks as if huge climate swings occurred in as little as decades—or even years. White and others think that, even in the warmer interglacial, a land could go from lush to desert, or temperate to icy,

## WEATHER EXTREMES

Increasingly, less stable weather conditions may be one of the chief perils of the climate change that appears to be under way. In a "greenhouse world," variability of climate would increase more than temperature.

**TASTE OF THE FUTURE?** New Yorkers brave gale-force winds and floods during a northeaster. Violent thunderstorms (left), tornadoes (top), floods, and droughts seem to be on the rise.

within a generation or less. One of the many things we still don't know is why we have been blessed with relatively steady conditions over the past ten millennia.

Although climate is the backdrop against which everything else happens, it's remarkable how little we understand of its workings. "Our scientific training ill

ECOLOGIST LEE KLINGER (left) and associate measure trace gases released by their homemade peat bog from northern Minnesota, now at the National Center for Atmospheric Research. Sphagnum, Klinger contends, takes over as peat bogs spread where trees once stood. Peat bogs worldwide form a huge carbon "sink"—drawing carbon dioxide out of the air and into the bog. Less $CO_2$ in the atmosphere correlates with cooler temperatures. Klinger speculates that bogs actually help bring on ice ages.

prepares us to deal with very complex systems like the atmosphere," says John Firor, formerly director of NCAR.

The more you look at climate, in fact, the more complicated it becomes. The flow of oceans and winds, changes in the earth's orbit, patterns of precipitation, different types of clouds, pollution— all these and more affect each other in mind-bogglingly complicated ways.

As I got a stronger sense of how much we don't know, I found it helpful to contemplate a favorite saying of Firor and other climate scientists: "The flap of a butterfly's wing in Argentina could affect the course of a tornado in Oklahoma."

As if the atmosphere weren't complicated enough, we ourselves have been throwing in wild cards. Again, John Firor: "It is now absolutely clear that the influence of people on the climate is as important as any of the other major influences....To ignore that is just plain wrong."

Perhaps the most significant influence of people on climate is the fact that we burn fossil fuels to power our economy. Burning coal and oil releases, among other things, carbon dioxide. We now pump about six billion tons of $CO_2$ from burning fossil fuels into the air each year.

$CO_2$, a by-product of animal respiration as well as burning trees and fossil fuels, efficiently traps infrared radiation—

NEW YORK CITY at dusk glitters with light from electric power plants that use oil and coal. Since the 1960s the concentration of heat-trapping $CO_2$ has been increasing at a growing rate, a fact that troubles most scientists.

Humans long considered the atmosphere and waters too big to harm. Now it is clear that pollution can kill lakes and forests hundreds of miles away. Scientists say the earth cannot indefinitely sustain such assaults.

heat from the sun that would otherwise bounce back out to space. So although it accounts for less than .04 percent of the atmosphere, $CO_2$ is a potent force that helps keep the planet habitable.

In fact, recent analysis of ice cores shows that the proportion of carbon dioxide in the atmosphere has varied naturally over the course of millennia. And the more $CO_2$, the higher the temperature.

None of the vagaries of atmospheric $CO_2$ in recent geological eras, however, compares with what is going on right now. Records from the climate observatory at Mauna Loa, Hawaii, show that a rapid increase in carbon dioxide in the atmosphere began around 200 years ago. $CO_2$ has risen more than 25 percent since the beginning of the Industrial Revolution.

In keeping with this, records of land and sea surface temperatures over the whole globe, averaged together, show that the 1980s was the warmest decade on record—and (continued on page 146)

SMOKE BILLOWING over the Amazon Basin (below) is emblematic of destruction occurring every day. Forests are vulnerable to accidental fires (left) and large-scale clearing for farmland, as well as for wood products.

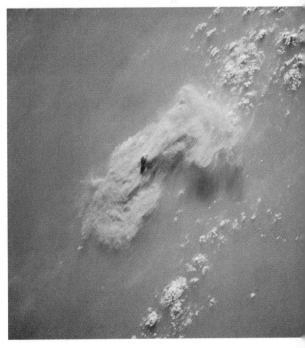

# CLIMATE CHANGE

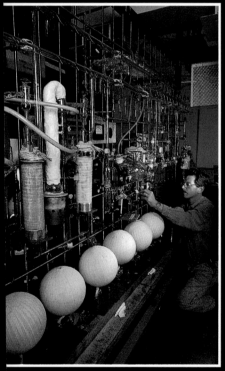

*Air samples collected in a vacuum flask (upper) help scientists analyze how fast $CO_2$ is rising (lower).*

Global climate has changed many times over the life of the earth. Glaciers have bulldozed prairies; sand dunes have blotted out lush wetlands; seasons have been far more extreme than they are today. For eons the world was so warm that even the poles were free of ice. Yet for the past 500 million years, anced on a knife-edge and can slide into one state or another within a human generation. All of civilization has blossomed during the last 10,000 years—apparently the steadiest climate the globe has known in hundreds of thousands of years. Yet the climate that humans take for granted—for agriculture, forests, way of life—could conceivably change very

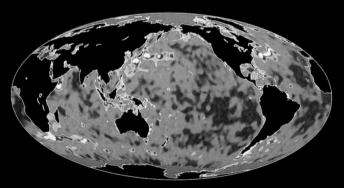

*Satellites facilitate monitoring of ocean currents (above). Computer models depict spread of polar ice (below).*

the earth has been spinning in and out of ice ages. Until a few years ago, climatologists thought such enormous changes took millennia. But new research indicates that climate is bal-

*Studying the effects of $CO_2$*

quickly. And it could change because of humankind. Analysis of ancient air trapped in polar glaciers shows that the more carbon dioxide in the atmosphere, the higher the temperature, and vice versa. As forests and fossil fuels burn, humans are adding $CO_2$ to the air at an unprecedented rate. For now the temperature of the globe appears to be rising— with weather unreliability increasing even more. Climatologists worry that, with human interference, the climate might flip into a different phase altogether—even tumble into an ice age.

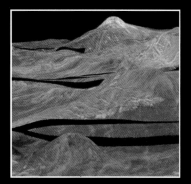

JUST ENOUGH carbon dioxide and other heat-trapping gases help keep the earth's temperature livable. The atmosphere of our neighbor Venus (above, in a computer-enhanced photograph) is far more $CO_2$-rich than ours: The planet's average temperature is 427˚C, much warmer than can be explained by its proximity to the sun.

*Iceberg blockades may disrupt ocean currents.*

1988, 1990, and 1991 the warmest years on record. In all, the average temperature of the world has risen by about half a degree Celsius over the past century.

Particularly troublesome are the class of chemicals called chlorofluorocarbons (CFCs)—some of which have 20 to 30 times the heat-trapping capacity of $CO_2$. Used extensively since the 1950s—primarily in refrigerators, air conditioners, and foam insulation—CFCs also are eating away the stratospheric ozone shield that protects us from the damaging ultraviolet rays of the sun.

Fortunately, the nations of the world banded together to phase out the production of CFCs by 2000. Nevertheless,

AN ATMOSPHERIC SCIENTIST readies an experiment (above) to be sent into the stratosphere over the Arctic. Balloons set to soar in Sweden (below) bear a payload of such tests to measure the concentration of ozone as well as pollutants in the upper atmosphere. The balloons sample clouds of ice crystals.

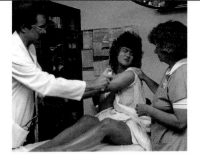

Above the earth hovers the ozone layer, which has shielded life from harmful ultraviolet rays for billions of years. Chemicals have eaten away at the layer—leading to a rise in sunburns (left) and an increase in skin cancer.

thin patches in the ozone umbrella over the northern and southern hemispheres persist, as does the ozone hole that appears each year over Antarctica. And because CFCs are long-lived, experts don't expect the ozone layer to recover for at least another century.

But climate change related to greenhouse gases is far more complicated—and controversial—than ozone thinning. That's partly because of gaps in our knowledge about how climate functions, and not the least because we can find no easy, benign substitute for fossil fuels. To reduce their output would require extensive changes in the way the world uses energy. So, although some diplomatic progress has been made, nations have been slow to take action.

Figuring out how climate change might affect the world is now on the cutting edge of climatology. What are these extra greenhouses gases actually doing? What roles do clouds, oceans, and different ecosystems play in warming and cooling the earth? What can previous climates tell us about what's likely to happen?

The quest for answers to these questions involves some of the brawniest computers yet devised. Climate modelers plug in equations that represent earth's size, rotation rate, seasons, topography and color of land, as well as the motions of water, air, and heat transfer. Then they vary factors—say, cloud cover or ocean currents or volcanic discharges—to see how the model says the system would react over time. The models concoct

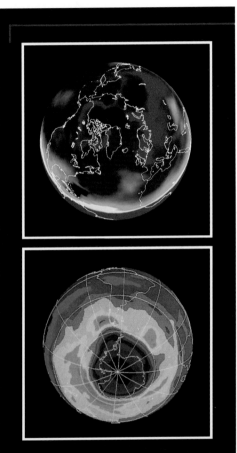

THE OZONE HOLE over Antarctica (lower) shows up as purple, pink, and, most severely, black in an image constructed from measurements taken by a satellite in 1990. The northern hemisphere (top), too, suffers an important though less dramatic fraying of its ozone shield. In the 1993 image, blue indicates a decrease of 5 to 15 percent over a decade; pink and purple, a loss of 15 to 25 percent. The black spot at the Arctic Circle represents inadequate data.

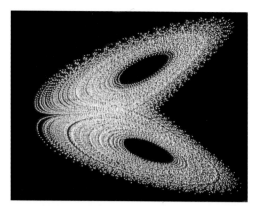

**COMPUTER-GENERATED FRACTAL demonstrates that minute changes of data can cause significant shifts in pattern from nearly identical points of departure. Such mathematical models imitate the forces of nature and suggest that turbulent, dynamic systems like weather can never be entirely predictable.**

coded, animated visions of the globe, showing snows coming and going, temperatures changing, oceans flowing.

Ask a climate modeler if he's making predictions, and he'll likely turn visibly uncomfortable. "We don't forecast anything; we can only do sensitivity experiments," says NCAR's Jerry Meehl. "Our models are still pretty crude; we need to keep improving them." Only in the past

few years has anyone built a model of ocean circulation; and experts are still at the early stages of coupling a model of the atmosphere to a model of the oceans for hundred-year experiments.

Meehl continues: "You don't take these model results literally, but as a warning. You look at it and say, 'What gets flagged here? What look like the biggest changes that may affect society?'"

Based on the best data and models available, scientists from the United Nations' Intergovernmental Panel on Climate Change (IPCC) agree that, given business as usual, the amount of carbon dioxide in the atmosphere will double due to human causes over the next century, increasing the global average temperature by 1.5° to 4.5°C.

While at a glance these numbers may look unimpressive, they represent a

## CLIMATE MODELING

Climatologists crunch numbers on mammoth computers to create visual models of possible climate states. They vary conditions, such as cloud cover and ocean currents, to project how climate might change over time.

world warmer than we've seen since the beginning of history. Add 5°C to today's global average and you get a globe hotter than it's been in the last 100,000 years. For another perspective: During the last ice age, when mile-high glaciers stretched from Long Island to Chicago to the Arctic, the average global temperature was only 5°C colder than now.

If the climate warms as the IPCC expects, swelling seas—not even counting the possibility of melting glaciers— would cause ocean levels to rise, drowning coastal cities and some entire island nations. Agricultural belts would shift dramatically, as not only temperatures but also rainfall patterns changed.

Moreover, with a rapidly changing climate, "practically every habitat on the planet will be put at risk," writes John Ryan, biodiversity specialist with Northwest Environment Watch of Seattle. Massive extinctions can be expected, since plants and animals are sensitive to climate and many ecosystems are already under stress from human incursions.

COMPUTER IMAGES (above) display visual models of gas and heat buildup on earth. Climate modelers use the data to understand how different factors might affect climate systems. Models are just now starting to capture the dynamics of ocean currents (left), which change more slowly than the atmosphere.

FOLLOWING PAGES: A classic fractal shows a basic pattern of nature. Today, scientists use fractals to analyze dynamic natural systems.

 nd yet the warming itself is not what many climatologists find most troubling: Extreme heat waves and cold fronts, hurricanes, floods, and droughts all appear to be on the rise. That's starting from the current yearly average of some 10,000 violent thunderstorms, 5,000 floods, more than 800 tornadoes,

and several hurricanes in the U. S. alone, according to the National Oceanic and Atmospheric Administration. Consider Hurricane Andrew, the Midwest floods of 1993, and the punishing winter of '94: If these are not products of global climate change, they give us a taste of what may be in store. (continued on page 152)

And, as climatologists say, climate always surprises us. One surprise that may bear on the future occurred about 11,000 years ago. The globe had just crawled out of the Ice Age and was nearing current temperatures. Suddenly, temperatures plunged and stayed low for 600 years.

**A SAN DIEGO FAMILY** displays what a typical household of four discards in a year. At left are 1,100 pounds of refuse to be recycled. At right are the 5,300 pounds of trash that find their way into landfills or incinerators (above).

Why the climatic flip-flop? Wallace Broecker of Columbia's Lamont-Doherty and George Denton of the University of Maine came up with a widely accepted hypothesis that requires a quick primer on a theory of ocean circulation:

A huge, looping ocean current—often called the conveyor belt—appears to circulate deep salty water from the tropical Pacific west to the Indian Ocean and the equatorial Atlantic. It flows north up the Gulf Stream to northern Europe. In the North Atlantic, the water releases its heat into the air, cools, and sinks. Eventually currents pull huge quantities of this dense, salty water back to the tropics, where it surfaces and heats up again.

This cycle seems to play a major role

## RECYCLING

Living lightly on the earth means using material in a thrifty manner—reducing how much paper, plastic, and metal is consumed; reusing containers and parts whenever possible; and recycling most of the rest.

in distributing heat around the globe. Turn off the Gulf Stream, and the whole world may be affected as the conveyor belt grinds to a halt. This seems to be what happened 11,000 years ago. Broecker and Denton proposed this possibility: Just before the episode, meltwater from the ice sheet thawing in North America was dammed up by glaciers. Suddenly, rafts of icebergs broke loose, enabling glacial meltwater to course into the North Atlantic. Being cold and fresh, this water sat atop the denser ocean water, forming a "freshwater lid."

Sea ice formed and, along with the icebergs, blocked escape of heat from below that, in turn, could not rise and warm the air currents; temperatures in the northern hemisphere, at least, plummeted. When the gush of meltwater ended, the conveyor belt tripped back on, and the Ice Age beat a full retreat—ending in just a few years.

Judging from the spikes in the ice-core record, something turned the conveyor belt off and on many times over the last 150,000 years. One new theory proposes a cycle of about 2,000 years: The Gulf Stream eventually warms the polar ice to the melting point. The ocean conveyor switches off. While the conveyor belt is stalled, temperatures plummet, until the melting stops, and the belt turns back on.

Another theory proposes armadas of icebergs ripping away from the North American continental ice sheet after several increasingly intense periods of warming. These iceberg armadas then sail off

into the North Atlantic—disrupting, once again, the ocean circulation pattern.

Other mechanisms may come into play as well. But it seems as if the North Atlantic is somehow key to climate—and vulnerable to disruption by freshwater flows. In a warming world, could melting ice in Greenland throw the switch and plunge us into an ice age? No one knows.

In fact, thinking about episodes like this "scares us," says Penn State's Richard

INSTEAD OF LANGUISHING in a landfill, old tires and aluminum cans (above) are reborn as insulation in the walls of an architect's office in Taos, New Mexico (below). Surprisingly, use of materials may affect the climate: It takes far more energy to produce a new tire, can, or bottle than it does to reuse or recycle one.

## RENEWABLE ENERGY

Earth is rich in energy sources that neither run out nor pollute—the wind, the sun, and earth's heat. Increasingly, humans are tapping this energy. Doing more work with less energy is also key to a renewable energy future.

Alley, who analyzed the temperature records. "We know that there are times when climate is very delicately poised. And we know that for the past 8,000 or 10,000 years, it hasn't flipped over. But we don't really understand climate well enough to say whether it's really stable or whether we're on thin ice."

There is much more to know. Even eminent climatologists sometimes suspect that, in the words of Gerard Bond, senior scientist at Lamont-Doherty, "There may be something really big that we're missing." Yet most believe we already know enough to begin to act.

As complicated as climate science is, "science is the easy part" compared to getting humans to agree on a course of action, says NOAA's Dan Albritton, veteran of two decades of international ozone treaty negotiations.

Albritton echoes the call of scientists around the world when he suggests, "You don't have to solve every question before doing anything: You can take initial action based on your initial understanding."

We know what curbing our uncontrolled planetary experiment will require: Moving from our economy based on fossil fuels to one built on solar and wind energy, using relatively clean natural gas as a bridge. It will mean designing our homes, factories, and appliances to use a fraction of the energy they now consume. It will mean developing a transportation system that takes advantage of energy-efficient public transportation and relies less on gasoline-burning automobiles. And it will mean helping developing nations leapfrog to the new energy systems.

**TECHNOLOGIES** like this rooftop solar panel—being installed in a Sri Lankan village—help developing nations leapfrog fossil-fuel-based energy systems to cleaner methods.

Are these ideas practical? Expensive? According to Amory Lovins, a leading energy expert and winner of a MacArthur "genius award," they're not only available and affordable—they're profitable. "Clouds of escaping $CO_2$ are really an indication of inefficiency in the marketplace," he says. Using superefficient technologies that are currently available, we "would save at least four-fifths of the electricity now used

**AMORY LOVINS, surrounded by new energy-efficient technologies, helped persuade utility companies to offer compact fluorescent bulbs and efficient shower heads. Behind Lovins: The GM Ultralite, the 100-mpg concept car that inspired him to make even more efficient designs.**

in the U. S.—and save money at the same time. We could also save at least four-fifths of the oil we now use—retrofitting for such efficiency is cheaper than drilling for more oil." Lovins has even designed a "super-car": an aerodynamic vehicle that he and interested auto manufacturers believe could get more than 150 miles per gallon.

"Increasing energy efficiency and moving to renewable energy are things we should be doing anyway—because they make economic and environmental sense, even if climate weren't a concern," concurs Nick Lenssen, energy analyst at the Worldwatch Institute. "Renewables—solar photovoltaics and wind turbines—are ready to be deployed on a large scale."

If we were to burn all the fossil fuels we think are left, we would increase the $CO_2$ in the atmosphere tenfold—com-pared to the doubling that troubles scientists today. There is another way—if we are wise enough to choose it. And future ice cores will show our descendants what action we took at this juncture.

Of course, some people find it comforting to take the long view. In the words of Sasha Madronich, "Whatever happens with $CO_2$ and all the other things we're pouring into the air, the planet will go back to its normal rhythms after we're gone. We're making things difficult for human life. But we're the ones who are here for the short term."

*Icelandic geothermal plant provides electricity and warm water.*

Although many uncertainties remain, a consensus of the world's scientists agrees that it is time to act to safeguard earth's climate. A key step, scientists believe, is to wean mankind away from carbon-rich fossil fuels. Technologies are already in place to harness the sun and wind, geothermal heat from the earth's core, and biomass, including waste from farms, trees, and other crops grown for

their energy content. And their cost is fast decreasing. The cost per kilowatt hour of wind power has fallen 75 percent in the

last decade, and that of generating electricity from sunshine by using solar photovoltaics has declined even more. The cost of building a wind farm is now comparable to that of a new coal-burning electricity plant—but saves the fuel, pollution, and carbon dioxide emissions. Hydropower—the energy in falling water—already provides one-fifth of the world's electricity. Biomass provides most of the energy for nearly half the world—some 2.5 billion people in developing countries. Wind supplies residential power to nearly a million Californians and could provide many countries

*Workers testing solar cells*

*Solar photovoltaic cells*

with at least one-fifth of their electricity. More than 200,000 homes in the developing world, and about 50,000 in Norway, are lit by solar cells, which convert sunlight into electricity. Engineers are vigorously trying to come up with cheap and reliable ways of storing energy from the sun and wind, as well as ways of transporting the energy during times when supply is high and demand is low.

Possibilities include using solar and wind power to produce hydrogen gas, clean-burning and easy to move. Energy experts estimate that 40 years from now, renewable energy sources could supply about half the world's energy. The potential is virtually infinite.

FOLLOWING PAGES: Solar photovoltaic cells in California light almost a million homes.

*World's largest wind power plant near San Francisco*

# OUTER SPACE

## by Dava Sobel

# OUTER SPACE

## 5

STARGAZING FROM THE FLAT ROOF of my garage, with its wide southern exposure of constellations, I see only the tip of the universe that professional astronomers examine. Unlike their giant telescopes and satellite-borne probes, my naked eye cannot penetrate the deceptive calm of the darkness—to see it erupt in the violent births and deaths of stars, the collisions of galaxies, and the buzz of radiation set loose at the beginning of time and space. Still, the night sky begs the same questions of me that researchers at the forefront of current astronomy pose. Like them, I want to know where the

The Milky Way arcs across the sparkling night sky in the Southern Hemisphere.
PRECEDING PAGES: A laser beams skyward at New Mexico's Starfire Optical Range.

## MAPPING THE UNIVERSE

New technologies allow astronomers to probe
unfathomable distances through astronomical time.
Recent studies challenge venerable theories and
reveal unanticipated patterns and structures.

galaxies go as they flee across the heavens. What gives rise to the stars? When did
the cosmos begin? Will it last forever? Are
we alone in the vastness of space? And
if not, who shares this universe with us?

These questions come out at night
with the stars to raise goose bumps on the
soul. As I wonder in the dark, I know that
instruments all over the world—and several more in orbit around it—are recording
the data that will provide answers at last
to fundamental mysteries of astronomy.

hat does the
universe look like?
The several thou-
sand stars that
I can see on a
clear, dark night consti-
tute a fraction of our Milky Way galaxy—
a swirling spiral of several hundred billion
stars, plus enough gas and dust to create
billions more. The mighty galaxy itself, however, is just one of a few dozen more or

less similar stellar assemblages in what
astronomers call the Local Group. These
fellow galaxies lie close enough to be
seen through an amateur's telescope.
Under ideal conditions, the soft, blurred
outline of the Andromeda galaxy, whose
light takes more than two million years to
reach the earth, reveals itself to the unaid-
ed eye like an intimation of infinity.

The Milky Way, the Andromeda
galaxy, the Magellanic Clouds, and other
members of the Local Group form but
one small neighborhood in a vast cluster
of galaxies. The Virgo cluster, the closest
cluster to us, may contain two thousand
individual galaxies. And clusters, in turn,
clump together to form superclusters that
are larger still. On the grandest scale
that astronomers can discern, thin sheets
of galaxies surround enormous voids in
space, forming a foam of cosmic bubbles.

Margaret Geller and John Huchra
of the Harvard-Smithsonian Center for
Astrophysics in Cambridge, Massachusetts,
who earned the nickname of "cosmic

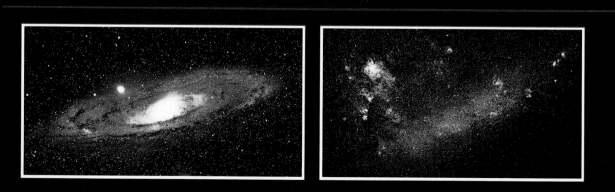

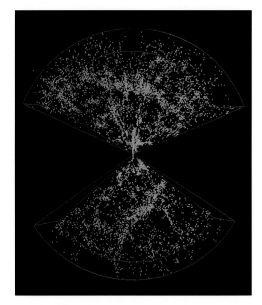

cartographers" for their efforts in mapping the lay of the nearby universe, were surprised by their own findings. They had not expected to see such a striking pattern of clustered galaxies and vast voids. Their three-dimensional, pie-slice picture of a region of space has become a modern icon as evocative as the double helix.

"Our main goal now," says Geller, "is to map the distant universe. At the depth that we'll go with our new telescope, we should get a detailed picture of what the universe was like five or six billion years ago."

Astronomers recently captured a detailed image of the infant universe as it appeared at least ten billion years back, soon after its explosive birth. No galaxies existed then. No stars had formed. There was only a fog of energetic particles, just beginning to sort themselves into the earliest outlines of structure. The Cosmic Background Explorer, or COBE, satellite charted these ripples in space-time by taking the most sensitive tempera-

**WEDGE-SHAPED SLICES** of the universe reveal large-scale structure first detected by Margaret Geller (opposite top, with student) and John Huchra. Viewed from earth's location at center, thousands of galaxies enclose vast voids.

**FOLLOWING PAGES:** Red box marks our sun's position in the Milky Way.

ture measurements ever achieved of the relic radiation from the original fireball. COBE maps reveal the ancient density differences that spawned today's patterns.

Surveys such as the one Geller and Huchra plan will let them observe the evolution of large-scale structure in the universe for the last *(continued on page 170)*

**THE LOCAL GROUP,** roughly 30 galaxies that cluster close to the Milky Way, spans three million light-years, very small compared to other clusters. Andromeda (far left), slightly larger than our own galaxy, surrounds its brilliant central bulge with glowing gas and spirals of young stars. The Large Magellanic Cloud (center), an irregular galaxy one-third the size of the Milky Way (left), glows so brightly that it has long been useful for celestial navigation.

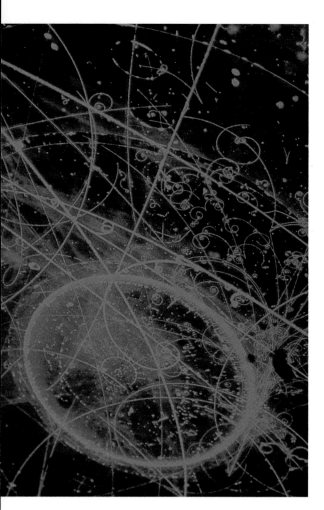

A BUBBLE CHAMBER (above) records the frenetic path of trapped neutrinos, elementary particles that lack mass and therefore are invisible. Collisions with other particles create the tracks. Astronomer Vera Rubin (opposite top) was among the first to study dark matter by recording faint light from the most distant parts of our galaxy on photographic plates.

Mexico State University, a participant in the Sloan survey. "As we look time and again at galaxies throughout the universe, we see lots of evidence that they have merged together, devoured each other, or disrupted one another through close encounters. Collisions between galaxies are very common—far more likely than the probability of two stars colliding within the Milky Way."

Even now, Burns points out, our own galaxy is tearing clumps of stars out of the Large Magellanic Cloud—enlarging itself by cannibalizing the smaller cluster.

What is the universe made of? Having pried open the heavens by examining every wavelength and nuance of light, astronomers now concede that the great bulk of the universe—perhaps

five to ten billion years. From the galactic history already gathered, astronomers hope to understand how the smooth universe shown by COBE became the artfully patterned universe we see today.

A new, ambitious project, the Sloan Digital Sky Survey Telescope, expects to image a hundred million galaxies over the next ten years. "It's a much more violent universe than I think we appreciated a few years ago," says Jack O. Burns of New

A THEORY OF DARK MATTER proposes the presence of invisible bodies called MACHOs, massive compact halo objects. Arrows point to a star in the Magellanic Cloud whose brightness changed during observation. Astronomers suggest that such an event indicates microlensing, a change in deflected light that occurs when a MACHO crosses the telescope's line of sight.

# DARK MATTER

Observations of visible bodies in our galaxy and others suggest massive gravitational influence from material that is not visible. At least 90 percent of all matter may be made of such particles referred to as dark matter.

90 to 99 percent of its mass—has escaped detection, by virtue of being dark and different from the familiar matter that shines.

Until now, scientists have dissected the electromagnetic radiation that comes to us from the stars and pulled information from each part of the electromagnetic spectrum. Visible light reveals the composition of the stars and the shapes of the galaxies. With the longer radio wavelengths, astronomers detect the birth memory of the universe and the death rattles of dying stars. Infrared astronomers can see the cosmic nests where new stars begin. Ultraviolet light exposes the hot exhalations that surround the stars; X rays mark the spots where black holes pepper the galaxies; and gamma rays stream from the hearts of active galactic nuclei.

But no such radiation emanates from the mysterious dark matter that must pervade the universe. It is invisible at every wavelength. Although observers can't see dark matter, they know it's there because of the way it makes stars and galaxies behave. Vera Rubin of the Carnegie Institution of Washington, D. C., was among the first to observe that the outermost stars in spiral galaxies travel at prodigious speeds. Yet they do not fly off into space. They seem trapped—gravitationally bound—in some formidable material. Astronomers call it dark matter.

Seeking dark matter in the Milky Way, Douglas Lin and his colleagues at the University of California, Santa Cruz, combined observations made over decades to estimate the total mass and extent of our galaxy. Lin's group reported in 1993 that the Milky Way weighs in at 600 billion solar masses—some five to ten times the combined weight of all its visible holdings. And while the diameter of the Milky Way had previously been shown to extend about 120,000 light-years across space, Lin says the true diameter more grandly spans 800,000 light-years. Aside from its role as galactic glue, *(continued on page 174)*

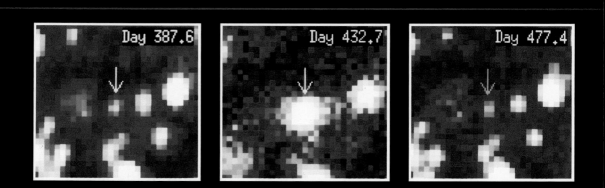

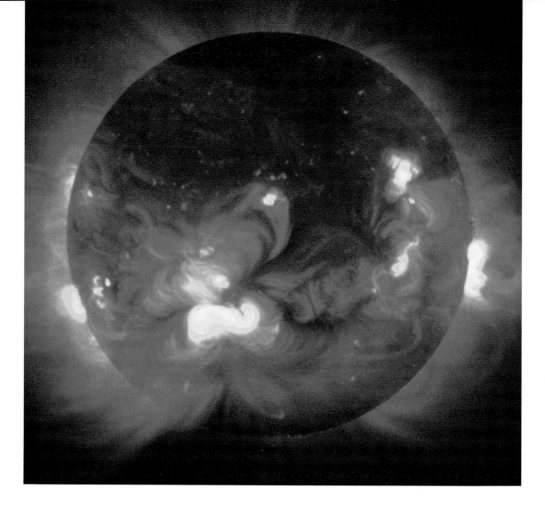

dark matter shoulders the responsibility for the fate of the universe at large. Given the fact that the universe is expanding, the galaxies may be destined to flee from each other eternally, becoming ever more distant. But, if the galaxies navigate through a sea of dark matter, then the universe may contain enough material to slow its outward expansion, and someday fall back together again. Thus the hunt is on to discover the nature and extent of this unknown substance.

Physicists and astronomers agree that some dark matter is made of ordinary material that takes the form of giant planets at least as large as Jupiter; of substandard stars, called brown dwarfs, whose nuclear fires never ignited; and of aged, exploded stars collapsed to the incredible densities of neutron stars and black holes.

Many theorists believe that some appreciable chunk, perhaps one-quarter, of dark matter consists of cosmic neutrinos—elusive subatomic particles. Other neutrinos have already been detected coming from the sun and from supernova explosions. Cosmic neutrinos, relics of the early universe, have not yet been detected, however.

"There are hundreds of these neutrinos in every cubic centimeter all around us," says David Caldwell, professor of physics at the University of California, Santa Barbara, and leader of INPAC—the Institute for Nuclear and Particle Astrophysics and Cosmology. "They have almost no mass, and they travel very, very slowly." Such properties make them virtually impossible to spot in space, but particle physicists using earthbound accelerators

## BIRTH AND DEATH OF STARS

Stars (left) begin as cool, gaseous globs enveloped in vast cocoons of interstellar dust. A cosmic tension created by the pull of its own gravity against the outward pressure of its exploding inner core defines each star's life span.

expect to establish their existence.

Other efforts focus on still more exotic breeds of dark matter, which should prove unlike any particles ever observed.

"The success of the big bang theory in accurately predicting the abundance of light matter in the universe sets clear limits on the amount of ordinary material in existence, so that leaves an awful lot of dark matter that has to be made of something else," says Caldwell. "The something else must be in the form of things that we simply don't know about at this point."

"It's not obvious how we're going to learn what the dark matter is," Vera Rubin comments, on the flurry of research activity unleashed by her findings in the 1970s. "It's not unusual for astronomers to stumble upon things that take us quite a while to understand. And so we don't worry about puzzles. We just keep approaching them from many different directions."

They hope to narrow the range of birth dates with findings from new techniques and instruments, including the recently repaired Hubble Space Telescope.

In this endeavor, the telescope honors Edwin Hubble, who noted in 1929 that galaxies all over the universe hurry away from us, outward bound. The farther they are, the faster they flee. Building on earlier work, Hubble could roughly clock their speeds by the *(continued on page 178)*

**THE SUN (opposite)** at five billion years of age is good for at least another five billion. Stars do burn out, however, and when they do, collapse to a density unlike any known substance. Similarly compressed, the Sears Tower (below) would become the size of a half-inch ball and weigh some ten billion pounds.

**FOLLOWING PAGES:** Exploding supernova (lower right) marks the death of a hot blue star.

**H**ow old is the universe? In the scientific version of genesis, the universe began as an almost infinitely hot, infinitely dense point no bigger than the head of a pin, before rushing headlong into the cosmic expansion that sowed the fields of the galaxies we see around us. Most astronomers enthusiastically support the big bang theory of creation. But they argue over the timing of the event—whether it took place as recently as 8 billion years ago or as far back as 20 billion.

## TELESCOPES

Indispensable to modern astronomy, telescopes have become exquisitely sensitive and complex devices capable of taking the pulse of unimaginably distant galaxies or probing the inferno-like surface of the sun.

down-shifting of their light into the lower frequency range of the electromagnetic spectrum, a phenomenon known as redshift. He began working on a scale to judge actual distances to calculate the expansion rate of the universe.

Taking up Hubble's challenge, astronomers today are still desperately seeking a reliable distance indicator—a milestone that will enable them to count off the megaparsecs across the warped landscape of space-time. (Astronomers who scale truly cosmic distances measure by megaparsecs. The gargantuan megaparsec is equal to 3,260,000 light-years.)

The most accurate distance indica-tor is a type of large, young star called a Cepheid variable. As Harvard astronomer Henrietta Leavitt showed in 1912, these stars grow brighter and dimmer in a predictable fashion: The longer their cycle of variation, the greater their true brightness. For example, a Cepheid variable with a cycle of ten days shines some two thousand times brighter than our sun. If the magnitude of a ten-day Cepheid appears fainter than that, then the faintness can safely be ascribed to distance, and the distance readily determined.

Hubble used the Cepheid stars to track the distance to the Andromeda galaxy, proving that it lay beyond the

UNOBSTRUCTED VISION from an optical telescope depends not only on clear skies and a compatible environment outside the observatory, but also on carefully controlled conditions inside the dome itself. The smallest temperature changes can distort star images. University of Washington scientists seeking ways to lessen interior turbulence conduct an experiment to test venting inside a structure. They submerge an acrylic model of a dome telescope (right) and use green dye to simulate air flow. A laser scanning beam illuminates the water tunnel.

Milky Way. But while they serve as "standard candles" that shine with the same intrinsic brightness wherever they occur, the Cepheids unfortunately fade from sight at huge extragalactic distances.

Before the repair mission that fixed the Hubble Space Telescope in 1993, the instrument could pick out Cepheid beacons in galaxies as far away as 12 million light-years. With its improved vision, the telescope has begun to discern Cepheids in galaxies at distances of 60 or 70 million light-years. This sixfold leap still doesn't carry astronomers out to the edge of the observable universe, billions of light-years hence; there, galaxies caught in the raging current called the Hubble flow approach speeds of true significance for settling the age debate. But the leap does enable astronomers to set a new zero

OPERATING AS ONE INSTRUMENT, the 27 radio-wave collectors of New Mexico's Very Large Array (left) search for signals too faint for optical detection. Protecting his eyes from a brilliance so dangerously close (opposite top), an astronomer at McMath Solar Telescope studies a reflected solar disk.

point on the cosmic measuring rod.

Wendy Freedman of the Carnegie Observatories in Pasadena, California, who heads the Hubble Space Telescope Key Project Team for the Extragalactic Distance Scale, has literally dusted off the Cepheid markers.

"Because they're young stars," Freedman explains, "they are found by astronomers near the dust and gas that formed them." Worried about that dust, Freedman and Barry Madore of Caltech showed how it could blanket a Cepheid's true brightness—making it appear dimmer and more distant. They also found a way around the dust by adding infrared observations to pictures obtained in visible light.

"When you make a correction for the dust," Freedman reports, "you find out that the distances to the Cepheids are smaller than we previously believed."

Recently, a team at the Carnegie Observatories has coupled Cepheids with brightness measurements of exploding stars, called supernovae, to arrive at a grand old universe of 15 billion years. The type of supernova used, known as a Type IA, outshines the *(continued on page 182)*

# MIRRORING THE SKY

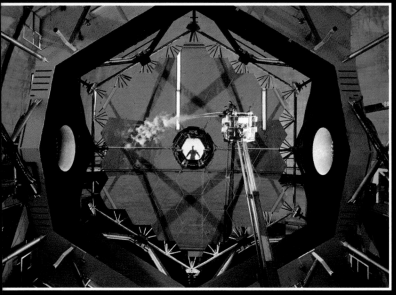

*Cleaning the giant mirror of Keck Telescope*

*Hexagonal mirrors*

image with two mirrors. The larger one, usually concave, gathers the light and brings the image to focus on a second, smaller mirror that deflects the image to an eyepiece lens. Numerous modifications have improved the design but not replaced it. To produce an accurate primary mirror, it is first necessary to meticulously grind and polish a large slab of glass until the desired concave shape is achieved. The glass is then coated with silver, aluminum, or another highly reflective substance. Light reflects off the surface rather than passing through it. The larger the mirror, the more light is intercepted and

to make. By the 20th century, most astronomers preferred using mirrors. The reflecting telescope, designed by Isaac Newton, produces its

The first telescopes were simple tubes with lenses fitted at each end. This instrument of Galileo revolutionized astronomy, but stargazers demanded larger telescopes and found that mammoth lenses were difficult and expensive

*Palomar Observatory's 200-inch reflecting telescope*

the sharper the image produced. Therefore the effort continues to produce the largest mirrors practicable. But huge mirrors—anything larger than 13 feet—can sag and distort the focus. Some observatories solve the problem by mounting composite mirrors, all aimed at the same focal point. One of the first

*Astronomer surveys observatory site at Cerro Paranal, Chile.*

giant multiple mirrors was built for the Keck Telescope, still the largest reflecting telescope in the world. The Keck mounts 36 hexagonal mirrors, each of which is six feet across and weighs about a thousand pounds. Other huge composites are planned for observatories in Chile and Japan.

*Keck Telescope, atop Mauna Kea, Hawaii, trains its eye outward.*

where Type IA beacons still beckon, though no Cepheids can be discerned.

Robert Kirshner at the Harvard-Smithsonian Center for Astrophysics, another devotee of supernova explosions, prefers the Type II variety that signals cataclysmic collapse of the most massive stars.

"Unlike the Type IA, which serves as a candle," Kirshner explains, "the Type II supernova is a real yardstick. It lets us measure distances in the universe in a way that is independent of the Cepheid method or any other technique."

For each supernova they study, Kirshner and his associates first assess the temperature, the brightness, and the expansion velocity of the stellar explosion. Then they use those figures to try to estimate the distance to the dead star.

"We're now getting out to distances where the galaxies are really part of the Hubble flow," Kirshner says excitedly. "This is where the big fish swim."

**THE HUBBLE SPACE TELESCOPE (above and top) docked with the U. S. space shuttle *Endeavour* in December 1993, while astronauts completed a delicate repair mission. Blurry vision corrected, the telescope once again floats free.**

Cepheids as a standard candle. If, as many astronomers assume, Type IA explosions always detonate with a characteristic charge, then their brightness belies their actual distance. The Carnegie team first spotted Cepheids and Type IA supernovae in the same galaxies and established the distances to those galaxies via the Cepheid scale. From there, they hop confidently to more distant galaxies,

ill the universe endure forever? To predict the future of the universe, astronomers look to its past—to the quantities of energy and substance it contained in the very first split seconds of its existence. For the ultimate fate of the universe depends on the mass of light and dark matter created in the big bang, and the rate at which all of it expands the boundaries of space-time.

**Telescopes equipped with powerful cameras circle the earth beyond atmospheric interference. Orbiting observatories facilitate research by conveying images of much higher resolution than those produced at ground-based observatories.**

If the amount of material is small and the expansion rate large, scientists suspect the universe will expand indefinitely—spreading, dimming, and chilling until its cinders cover an incalculable volume of space. If, however, the universe has great substance and a relatively meager rate of expansion, it will reach the point where the braking action of its own mass will pull back in on itself—reheating as it collapses all the way down to a "big crunch," as dense and hot as the big bang.

A third possibility is outlined in the most popular model of the big bang theory, called the inflationary universe. First proposed by Alan Guth of the Massachusetts Institute of Technology, this idea envisions a universe that suddenly mushroomed as it got under way, so that one minuscule region expanded exponentially in less than a few trillionths of a second, putting the farthest reaches forever out of our sight. This theory requires the universe neither to be headed for infinite dispersion nor to be massive enough to rebound, but to be poised precisely between these two

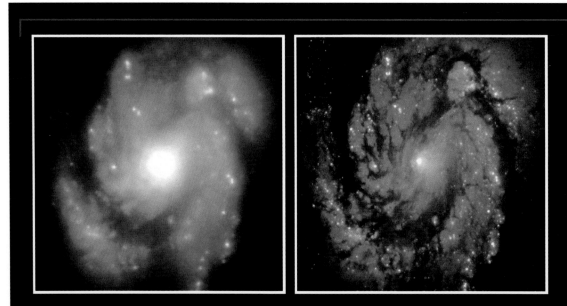

**IMAGES CONVEYED BY THE HUBBLE SPACE TELESCOPE explore a distant bright galaxy known simply as M100. An early view (left) produced by Hubble's Wide Field and Planetary Camera (WFPC) surpassed images possible at ground level but nevertheless appeared disappointingly blurred. A later image (right), produced after the 1993 repair mission, sharpens the focus and resolves detail in the spiral structure, making possible a first look at dust clouds, gases, and individual stars deep within the galaxy's core. According to a Hubble scientist: "These corrected images are as perfect as engineering can achieve and physics will allow."**

extremes. This "flat" universe embodies just enough mass to continue expanding forever at a constantly decelerating rate.

Many astronomers are assessing the density of the universe and trying to see if it matches the so-called critical density required for flatness and eternal expansion. In this realm, however, inherent limits on observation may leave several big questions unanswered.

"I have a feeling that the inflationary universe model will still not be confirmed or refuted over the next decade," observes astronomer and author Alan Lightman of MIT. "We know the universe is close to flat, and the observations that would show it to be exactly flat are extremely difficult. Furthermore, a lot of the interesting activity in cosmology has been pushed back earlier than the first nanosecond, during a period of time that we will not be able to probe with telescopes or with particle accelerators on earth. Theory has gotten very far ahead of experiment. But, gradually, the inflationary universe model may simply become

accepted as a standard part of the assumptions that cosmologists make."

**W**ho shares the cosmos with us? The events that populated the earth, scientists say, were wonderful but not miraculous. Similar scenarios could have taken place on countless other planets of other stars. Although astronomers have not yet identified a planet of another star, they have seen great disks of

RADIO WAVES

MICROWAVES

INFRARED

VISIBLE LIGHT

ULTRAVIOLET

X RAYS

GAMMA RAYS

**VISIBLE LIGHT, viewed by optical telescopes (above), represents only a small fraction of the energy transmitted to us from the sun and from remote parts of the universe. Electronic oscillation within each band determines its wavelength.**

Balloons, rockets, and earth-orbiting satellites now make it possible to monitor electromagnetic radiation emitted with invisible wavelengths. Each band of energy holds clues to the nature and history of the universe.

material surrounding distant points of light the way the solar system circles the sun. They believe that many, if not most, stars enjoy the company of planets, and that a sizable fraction of those planets harbor intelligent life.

Radio astronomer Frank Drake of the University of California, Santa Cruz, conducted the first search for alien civilizations more than 30 years ago at West Virginia's National Radio Astronomy Observatory. Drake also devised an equation to estimate the number of intelligent civilizations in the Milky Way. He now estimates that some ten thousand exist and could be detected via the radio signals leaking from their home planets into space. Earth has been broadcasting news of human existence since the advent of widespread television in the 1940s. An ever widening sphere that now extends some 50 light-years beyond the solar system could alert attentive aliens to the facts of our lives.

"Personally," says Drake, "I find nothing more tantalizing than the thought that radio messages from alien civilizations in space are passing through our offices and homes, right now, like a whisper we can't quite hear."

**VIEWED BY EQUIPMENT** sensitive to photons above and below visible light, these images, bisected by the Milky Way, reveal our galaxy. Radio waves show galactic structure. Microwaves compare local temperatures to the average for the universe. Infrared shows the Milky Way's central bulge. Ultraviolet maps hot nearby stars. A glow of X rays fills the sky. Gamma rays show where cosmic rays collide with gas.

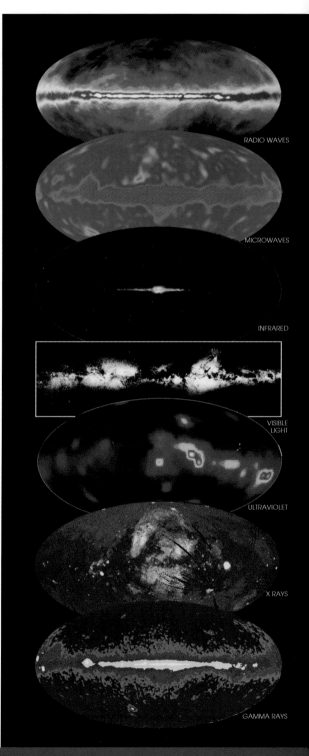

RADIO WAVES

MICROWAVES

INFRARED

VISIBLE LIGHT

ULTRAVIOLET

X RAYS

GAMMA RAYS

**STANLEY MILLER re-creates his experiment of the 1950s, which suggested that we are not alone in space. Gases in the bulb simulate earth's atmosphere shortly after its creation. An artificial bolt produces amino acids, without which life as we know it could not exist.**

The National Aeronautics and Space Administration launched an ambitious program in the search for extraterrestrial intelligence, known as SETI, in October 1992. This all-sky survey and search of some one thousand sunlike stars was to continue for ten years, utilizing a network of radio telescopes in Puerto Rico, California, Australia, and Europe. Within one year, however, opposition arose, and Congress cut the project's funds, to NASA's chagrin.

Interest in the discovery of extrater-restrial intelligence runs so high among the public at large, however, that NASA's SETI project and several other long-running search endeavors continue to operate on money from private donations. Renamed Project Phoenix, the former NASA SETI mission has attracted millions of dollars in grants from industrial leaders. Another ongoing effort, called Project META— Megachannel Extraterrestrial Assay— receives its support from the Planetary Society, a grassroots organization of space enthusiasts founded by astronomers Carl Sagan of Cornell University and Bruce Murray of the California Institute of Technology. The members include more than a hundred thousand individuals willing to spend their money to explore the

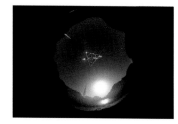

For more than 30 years astronomers have sifted through radio signals from our galaxy and beyond hoping to detect intelligent life. Radio telescope at Puerto Rico's Arecibo (left) scans the skies for faint emissions.

intriguing possibility that we are not alone.

Sagan and colleagues recently used remote sensing equipment aboard the Galileo spacecraft to examine earth from an extraterrestrial perspective. The satellite's course took it over the South Pole and Australia, where it failed to record the nighttime lights of the great metropolitan areas or the quilt-block patterns of cultivated fields. Yet Sagan's group did detect signs of life. They saw an atmosphere rich in oxygen—a true anomaly among the known planets. They watched the land surfaces glow brightly at some wavelengths and dim out at others, in a signature that clashed with the rocks and soils of lifeless worlds. And they came upon telltale pulses of radio energy, concentrated in narrow bandwidths, that appeared unique in all the universe.

"Galileo found such profound departures from equilibrium," Sagan reports, "that the presence of life seems the most probable cause." Thus, the earth shows itself to be inhabited by creatures that have mastered technology.

Does intelligent life exist elsewhere in the galaxy? The detection of radio signals from a single extraterrestrial source would answer this simple question in the affirmative. But a "no" answer is unattainable. Even if grandiose searches operate at full tilt for decades and find no substantial evidence, the possibility remains that signals still lurk in unexamined regions, are blocked by galactic interference, or that extraterrestrials—through feats of technology—no longer broadcast radio waves

into space. The *failure* to find other forms of life offers no proof of our solitude.

The constellations pass with the hours over my rooftop, as the spinning planet wheels the heavens around me. I savor the illusion of being at the hub of the universe—even though I know that centuries of progress have ousted the earth from the center of the solar system, banished the sun to the outskirts of the galaxy, and exposed the material of which our stars and our bodies are made to be not just dust, but only a film of dust on the face of the vast dark universe.

Instead of feeling reduced to insignificance by these revelations, astronomers read in them a dare to fathom all the rest.

**WAITING THROUGH YEARS of long nights, sifting through the eternal flow of data, astronomers patiently investigate any distant signal of unexplained origin.**

**FOLLOWING PAGES: A directional device aims the telescope at Arecibo toward a specific star, listening for an unfamiliar pulse.**

# AS FAR AS EYE CAN SEE

*An Epilogue
by Arthur C. Clarke*

FOR MANY YEARS I have amused myself
by showing two small artifacts to visi-
tors, and asking them to guess their
origins. One is a bent, square-sectioned
copper nail. The other is more mysterious: It's a piece of black, charred
material with a honeycomb structure.
I usually give my victims a clue: These
are relics from two of mankind's most
famous voyages of discovery. Before
frustration becomes unbearable,
I reveal their origins. The nail is from
H.M.S. *Bounty* and was given to me by
the National Geographic Society's Luis
Marden, who discovered the wreck
where the mutineers had burned and
sunk it off Pitcairn Island. And the piece

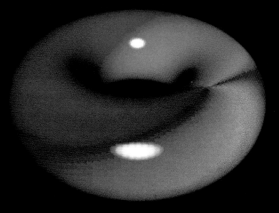

*Computer graphic of a torus*

of carbonized honeycomb comes from the heat shield that protected the crew of Apollo 8, when they re-entered earth's atmosphere after the first circumnavigation of the moon.

Few objects demonstrate more dramatically the amazing advances that technology has made in only 180 years; by comparison, all of previous history seems merely a prelude. Can this rate of change continue, or will the discoveries gradually subside, so that mankind eventually attains a steady-state future?

Though this may be true in the long run — perhaps a few thousand years — I suspect that we are nowhere near that comfortable but boring state of affairs. Anyone who believes otherwise should recall the proposal, at the end of the last century, that the U. S. Patent Office be abolished since nothing important was left to invent.

As far as discovery is concerned, we can be quite sure that nature still has many more surprises up her capacious sleeve. The great biologist J. B. S. Haldane summed it up perfectly when he said that the universe is not only stranger than we imagine — it is stranger than we *can* imagine. Who would ever have dreamed of pulsars — stars spinning hundreds of times a second, sweeping the sky with radio beams; of gamma-ray bursters — cosmic enigmas that can switch on the power of a billion suns in a matter of minutes; of neutron stars, on whose surface a man would be instantly flattened into a microscopically thin film? Not to mention black holes, which fold space around them-

selves so that nothing — not even light itself — can escape.

These examples should warn us how difficult — no, impossible — it is to anticipate the most exciting technologies of the future because there may be powers and forces in nature that are still undiscovered and thus unexploited. Not even Ben Franklin could have dreamed how we would tame lightning to perform a myriad of household tasks. Edison himself would have been completely baffled by a laser, a pocket calculator, a hologram, or a compact disk. And if he had looked at a silicon chip through a microscope, he would not have believed his eyes. This tiny marvel, which has revolutionized our world, is the best proof of what I call Clarke's Third Law: "Any sufficiently advanced technology is indistinguishable from magic." Well, here are a few possible magics that may also shape the future.

The exploration of space has been the main interest of my life, and I count myself extremely fortunate to have shaken hands with the first man to orbit the earth, the first to do a space walk, and the first to set foot on the moon. Unfortunately, it costs at least ten thousand dollars *per kilogram* to send anything into orbit, and although improvements in rocket technology may allow limited space tourism during the next century, only the rich will be able to afford holidays on the moon. Yet, incredibly, the cost of a moon trip, in terms of energy, is about a hundred dollars per person. Rockets are not only inefficient but environmentally deplorable and potentially dangerous. There must be a better way.

There is, at least on paper. Don't laugh: It's an elevator. In 1960 a Russian engineer, Yuri Artsutanov, pointed out that it would be theoretically possible to lower a cable from a satellite hovering in geostationary orbit to the earth 36,000 kilometers below. On this foundation one could construct what Artsutanov called a "celestial funicular." And then payloads and passengers could be lifted into space at comfortable velocities and negligible cost. In fact, the energy cost would be almost zero because payloads descending to earth would lift outgoing ones.

Of course, building the space elevator would require materials of enormous tensile strength, and the finest steels would be inadequate for the job. However, in 1993 scientists at Rice University in Houston, Texas, discovered a new form of carbon, named after the famous engineer Buckminster Fuller. This, they proudly announced, would make it possible to construct a space elevator. I was delighted to hear this, as I had already assumed the use of carbon, or diamond fiber, in my novel *The Fountains of Paradise.*

Later that same year I was even more delighted to receive a letter from astronaut Jeffrey Hoffman, payload commander on the STS-46 *Atlantis* mission. This shuttle carried a satellite that was to be lowered from orbit on a 20-kilometer-long tether. It also carried *The Fountains of Paradise,* which all the crew signed. Jeff Hoffman's covering letter remarks: "We all read your book while we were preparing for our mission. It represents the ultimate use of tethers in space, and we felt privileged to take part in the first step on the way to this goal."

Though the space elevator could give cheap access to space, the rocket would still be needed for travel onward to the moon and planets and for the return to earth. But will the rocket itself ever be superseded by something better?

Science fiction writers have long dreamed of a mythical "space drive" that would allow us to go racing round the universe — or at least the solar system — without the rocket's noise, expense, and danger. Until now, this has been pure fantasy, and it may always be so. However, recent theoretical studies — based on some ideas put forward by the great Russian physicist and human rights campaigner Andrei Sakharov — hint that some control may indeed be possible over the mysterious forces of gravity and inertia.

# INDEX

**Boldface** indicates illustrations.

## NOTES ON CONTRIBUTORS

Science fiction writer ARTHUR C. CLARKE has published more than 70 books. Known for conjuring daring images of the future, the British-born Clarke has lived in Sri Lanka for the past 30 years. In 1994 he was nominated for the Nobel Peace Prize for his 1945 theory of communications satellites.

CAROLE DOUGLIS is an environmental writer whose work has appeared in Atlantic Monthly, Omni, Psychology Today, Harpers, World Watch, and numerous publications of the National Geographic Society. She lives in Washington, D. C.

ROBERT FRIEDEL is a professor of technology and history at the University of Maryland, College Park. His research and writing have resulted in books on the history of plastics, electric lighting, the zipper, and materials in American life.

STEPHEN S. HALL writes frequently about molecular biology and the history of science. His most recent book, Mapping the Next Millennium, is about the convergence of new technologies with new forms of cartography. He is currently working on a history of immunotherapy.

Free-lance space and science photographer ROGER H. RESSMEYER resides in Sag Harbor, New York. A graduate of Yale, Ressmeyer began his career as a portrait photographer. A frequent contributor to NATIONAL GEOGRAPHIC magazine, he has authored two picture books about exploring the universe.

Neurologist and neuropsychiatrist RICHARD RESTAK has written ten books on the brain, his latest entitled The Modular Brain. A clinical professor of neurology at George Washington University, Restak lectures frequently on the brain. He lives and practices in Washington, D. C.

DAVA SOBEL, former science reporter for the New York Times, writes about astronomy for Omni, Discover, Life, Audubon, and Harvard magazines. She is co-author, with astronomer Frank Drake, of Is Anyone Out There? The Scientific Search for Extraterrestrial Intelligence. Her book A Family Portrait of the Planets is due out soon.

WALTER SULLIVAN, former science editor for the New York Times, has covered many of the major science events of the last decades. In addition to his news coverage, he has written several books on science-related subjects. In 1994 he was elected to the American Academy of Arts and Sciences.

## ACKNOWLEDGMENTS

The Book Division wishes to thank the individuals, groups, and organizations named or quoted in the text. In addition, we are grateful for the assistance of Paul A. Hyslop, Thomas Kelsall, Len Lessin, Kurt Rasmussen, and Jain Ressmeyer.

## ADDITIONAL READING

The reader may wish to consult the National Geographic Index for related articles and books. The following titles may also be of interest: John Finchar, The Brain: Mystery of Matter and Mind; John Firor, The Changing Atmosphere: A Global Challenge; Tom Forester, The Materials Revolution; Stephen S. Hall, Mapping the Next Millennium; Joseph Levine and David Susuki, The Secret of Life; Dennis Overbye, Lonely Hearts of the Cosmos; Richard Restak, The Brain; Stephen H. Schneider, Global Warming: Are We Entering the Greenhouse Century?; Raoul Smith, Artificial Intelligence.

Abbreviations for terms appearing below: (t)-top; (b)-bottom; (l)-left; (r)-right; (c)-center; PA-Peter Arnold, Inc.; TIB-The Image Bank; PR-Photo Researchers; SPL/PR-Science Photo Library/Photo Researchers.

**COVER:** Richard Wahlstrom/TIB.

**FRONT MATTER:** 1, Richard Wahlstrom/TIB. 2-3, Dominique Sarraute/TIB. 4-5, Roger H. Ressmeyer. 6-7, Hank Morgan. 8-9, J. J. Hester, Arizona State Univ., and NASA. 10-11, Roger H. Ressmeyer, Starlight. 13 (t), Joe McNally/TIB; (ct); Jim Richardson/Westlight; (c), Dominique Sarraute/TIB; (cb); Art Matrix/Rainbow; (b), Roger H. Ressmeyer, Starlight.

**INTRODUCTION:** 14, M. Simpson/FPG Int'l. 16, Bruce Frisch/PR. 17 (t), Peter Menzel; (b), Charles O'Rear/Westlight. 18, P. Motta/Dept. of Anatomy/Univ. "La Sapienza", Rome/SPL/PR. 19, Philip A. Harrington/The Stock Market.

**CHAPTER 1 - THE BRAIN:** 20-21, Joe McNally/TIB. 22, Manfred Kage/PA. 24-25, Mallinckrodt Institute of Radiology/Washington Univ. School of Medicine/Michael W. Vannier/digitally enhanced by Roger H. Ressmeyer. 25, Roger Tully/Tony Stone Images. 26 (t), Terry Qing/FPG Int'l; (b), Hank Morgan/Rainbow. 27 (tl&tr), Roger H. Ressmeyer; (b), Thomas Mayer/Black Star. 28-33 (all), Roger H. Ressmeyer. 34, Marcus E. Raichle/Washington Univ. School of Medicine; (b), Roger H. Ressmeyer. 35, Marcus E. Raichle/Washington Univ. School of Medicine. 36 (t), David M. Porter/F-Stock, Inc.; (tl & bl), Philippe Plailly/SPL/PR; (br), Hank Morgan; 37 (t), Louie Psihoyos; (b-all), Hank Morgan. 38-39, Ted Spiegel/Science Source/PR. 40, Karen Kasmauski. 41 (tl), Alain Soldeville/Rapho/Black Star; (tr), Mel Digiacomo/TIB; (b), Lawrence Fried/TIB. 42 (t) Jan Halaska/PR; (b), Horst Ebersberg/Rapho/PR. 43 (t), Leonard Lessin/PA; (b), Martin Rogers/Tony Stone Images. 44 (tl & tr) unknown, NGS Lab; (cl), Roger H. Ressmeyer, Starlight; (c&cr), D. Miller/PA; (b), Laura Dwight/PA. 45, Lynn Johnson/Black Star. 46 (tl), Roger H. Ressmeyer; (tr), John Cole/Impact Photos; (b), Bruce Stephens/Impact Photos. 47, Peter Van Mier & Jeff Lichtman/Washington Univ. Medical School/digitally enhanced by Roger H. Ressmeyer; (b), Peter Menzel. 48 (both), Roger H. Ressmeyer. 49 (tl & tr), Hank Morgan/Rainbow; (b), Roger H. Ressmeyer. 50 (t), Dan McCoy/Rainbow; (c&b), Peter Menzel. 51 (t), Francesco Ruggeri/TIB; (b), Michael Melford/TIB. 52-53, Rick Friedman/Black Star.

**CHAPTER 2 - NEW BIOLOGY:** 54-55, Matt Meadows/PA. 56, Peter Menzel. 58 (tl), Ed Reschke/PA; (tr), T. Ried & D. Ward/PA; (b), Steve Winter/Black Star. 59 (t), James Holmes,Cellmark Diagnostics/SPL/PR; (b), Douglas Struthers/Tony Stone Images. 60, Francis Leroy, Biocosmos/SPL/PR. 61 (t), David M. Porter/F-Stock, Inc.; (c), Jon Gordon/Phototake; (b), Hank Morgan. 62, Roger H. Ressmeyer. 63 (t), Donal Philby/FPG International; (c), Peter Menzel; (b), Roger H. Ressmeyer. 64-65, Peter Menzel. 66 (tl), Brownie Harris/The Stock Market; (tr), Günter Beer/Visum; (bl), Sepp Seitz/Woodfin Camp & Assoc. 67 (t), George Olson/Woodfin Camp & Assoc.; (c), Chris Johns; (b), Peter Menzel. 68, Art Montes De Oca/FPG Int'l. 69 (t), Phil Matt; (c), Gerd Ludwig/Woodfin Camp & Assoc.; (b), Jonathan Craymer/Rex Features. 70 (l), Karl Hartmann/Sachs/Phototake; (r), Louie Psihoyos/Matrix. 71 (t), Peter Menzel; (b), Hank Morgan/Rainbow. 72 (t), Dennis Brack/Black Star; (l), Mark R. Holmes, Univ. of Utah, Dept. of Human Genetics. 72-73, Peter Menzel. 73, Peter Menzel. 74, Larry Mulvehill/PR. 75 (both), Peter Menzel. 76-77, Jim Richardson/Westlight. 78 (t), S. Walkley/PA; (b), Arnold Zann/Black Star. 79, Nick Kelsh. 80, Ted Spiegel. 81 (t), Roger H. Ressmeyer; (b), Genentech/UCSF Midas Plus/Hynes/de Vos/Andow/digitally enhanced by Roger H. Ressmeyer. 82 (t), Matt Meadows/PA; (c), David Scharf/PA; (b), CNRI/SPL/PR. 83 (t), Lennart Nilsson, Boehringer Ingelheim Int'l GmbH; (c), Lennart Nilsson; (b-1), Tektoff-RM/CNRI/SPL/PR; (b-2&3), Moredun Animal Health Ltd/SPL/PR; (b-4), CNRI/SPL/PR; (b-5), Gopal Murti/SPL/PR. 84 (l), Al Lamme/Phototake; (r), NIH/Custom Medical Stock Photo. 85, Dennis Galloway. 86 (t), T. Stephan/Das Fotoarchiv/Black Star; (b), Lynn Johnson/Black Star; (b), Charles Daguet/Pasteur Institute Petit Format/Science Source/PR. 87 (t), Richard Falco/Black Star; (bl), John Harrington/Black Star; (br), Peter Menzel. 88-89, Yoav-Simon/Phototake.

**CHAPTER 3 - NEW MATERIALS:** 90-91, Dominique Sarraute/TIB. 92, Charles O'Rear/Westlight. 94 (l), Digital Art/Westlight; (r), Mike & Carol Werner/Comstock. 95 (t), Alfred Pasieka/SPL/PR; (b), Manfred Kage/PA. 96 (t), Michael W. Davidson/PR; (cl), Melvin L. Prueitt, data from Fred Mueller, Los Alamos National Laboratory; (tr), Scott Camazine/PR; (br), Biosym Technologies Inc./SPL/PR. 97 (all), Charles O'Rear/Westlight. 98-99, Geoff Tompkinson/SPL/PR. 100 (t), James Stoots/Univ. of CA, Lawrence Livermore National Laboratory; (b-all), Melvin L. Prueitt. 101 (t-all), Hank Morgan/PR; (bl), Erich Schrempp/PR; (br), Charles O'Rear/Westlight. 102 (tl&tr), Charles O'Rear/Westlight; (bl) Melvin L. Prueitt, data from Victor F. Zackay, Los Alamos National Laboratory; (br), Frank W. Gayle and Alexander J. Shapiro, National Institute of Standards and Technology. 103 (l&tr), James L. Amos; (br), Charles O'Rear/Westlight. 104, Melvin L. Prueitt; data from Paul G. Riewald, Du Pont, and from Fionni Dowell, Los Alamos National Laboratory. 105 (t), Stephanie Stokes/The Stock Market; (b), Charles O'Rear/Westlight. 106-107, Charles O'Rear/Westlight. 108 (t), Charles O'Rear/Westlight; (b), Roger H. Ressmeyer. 109 (t), Dan McCoy/Rainbow; (b), Roger H. Ressmeyer. 110 (t), Peter Menzel; (bl), Roger H. Ressmeyer; (br), Bernard Roussel/TIB. 111 (tl), Charles O'Rear/Westlight; (tr), Comstock; (cr), Siu Biomed Comm/Custom Medical Stock Photo; (b), Roger H. Ressmeyer. 112 (t), Astrid & Hans

Frieder Michler/SPL/PR; (c), Melvin L. Prueitt; (b), Roger H. Ressmeyer, Starlight. 113 (both), Roger H. Ressmeyer. 114 (t), IBM Corporation, Research Division, Almaden Research Center; (c), Peter Menzel; (b), Philippe Plailly/SPL/PR. 115 (l), Charles O'Rear/Westlight; (c&r), IBM Research/PA. 116, Karlsruhe Nuclear Research Center. 117 (tl), Peter Menzel; (tr&cr), K. Eric Drexler, Inst. for Molecular Manufacturing and Ralph Merkle, Xerox, Palo Alto Research Center; (cl&bl&br), Peter Menzel. 118 (t), Derek Redfearn/TIB; (b-all), Leo Barish/Albany Int'l Research Co. 119 (tl), Dianne Blell/PA; (tr), Robert A. Edahl, Jr., Langley Research Center/NASA; (b), Dan McCoy/Rainbow. 120 (t), Roger H. Ressmeyer; (c), Lawrence Livermore National Laboratory/SPL/PR; (b-all), Cargill. 121, Roger H. Ressmeyer. 122-123, Roger H. Ressmeyer, Starlight.

**CHAPTER 4 - CLIMATE:** 124-125, NASA/Comstock. 126, Los Alamos National Laboratory. 128 (t), Fred Ward/Black Star; (b), Frank Rossotto/The Stock Market. 129 (l), Roger H. Ressmeyer, Starlight; (r), Stephanie Maze. 130 (t), CSIRO/SPL/PR; (b), Peter Menzel. 131 (l), Doug Allan; (tr), CSIRO/SPL/PR; (br), George F. Mobley. 132-133, Roger H. Ressmeyer. 134 (l), Harvey Weiss; (r), Nathan Benn. 135 (tl), James L. Stanfield; (tr), Peter de Menocal/Lamont-Doherty Geological Observatory of Columbia Univ.; (bl), Reza. 136, Bill Eldred. 137 (l), Animals Animals/Thomas Long; (r), Robert M. Carey, NOAA/SPL/PR. 138-139, Alberto Garcia/Saba. 140, NASA/Science Source/PR. 140-141, Ray Nelson/Phototake. 141 (t), Gene Moore/Phototake; (c), Newsday/Ken Sawchuk. 142 (t), Roger H. Ressmeyer; (b), Nicholas DeVore III/Photographers Aspen. 143 (t), Oddo & Sinibaldi/The Stock Market; (bl), Paul Chesley/Photographers Aspen; (br), Frank P. Rossotto/The Stock Market. 144 (tl), Hank Morgan/PR; (bl), Hank Morgan/Rainbow; (tr), NASA/SPL/PR; (br), NCAR. 145 (tl), Roger H. Ressmeyer; (tr), NASA/Mark Marten/Science Source/PR; (b), Bryan and Cherry Alexander. 146 (both), Philippe Plailly. 147 (tl), James A. Sugar/Black Star; (tr), NASA/Science Source/PR; (b), NASA/SPL/PR. 148 (t), E. Lorenz/PA; (b), Los Alamos National Laboratory. 149, Hank Morgan/Rainbow. 150-151, Art Matrix/Rainbow. 152 (t), Roger H. Ressmeyer; (b), José Azel. 153 (both), Roger H. Ressmeyer. 154 (t), James A. Sugar/Black Star; (b), Solar Electric Light Fund. 155, Roger H. Ressmeyer. 156 (t), Roger H. Ressmeyer, Starlight; (b), Harald Sund/TIB. 157 (t), Peter Menzel/Stock Boston; (b), Peter Menzel. 158-159, Dan McCoy/Rainbow.

**CHAPTER 5 - OUTER SPACE:** 160-161, Roger H. Ressmeyer, Starlight. 162, Roger H. Ressmeyer. 164 (t), Jim Harrison; (bl) Tony Hallas/SPL/PR; (br), Royal Observatory, Edinburgh/SPL/PR. 165 (t), Margaret J. Geller, John P. Huchra, Luis A. N. da Costa, and Emilio E. Falco, Smithsonian Astrophysical Observatory; (b), Roger H. Ressmeyer, Starlight. 166-167, Joe Tucciarone. 168 (t), Roger H. Ressmeyer; (bl), David Hardy/SPL/PR; (br), Mehau Kulyk/SPL/PR. 169 (t), J. J. Hester/Arizona State Univ./digitally enhanced by Roger H. Ressmeyer; (b), NASA, Cobe Science Team. 170, Dan McCoy/Rainbow. 171 (t), Patricia Lanza, courtesy KCET, Los Angeles; (b-all), Charles Alcock, Macho Collaboration/SPL/PR. 172 (l), David Hardy/SPL/PR, (r), Tony Craddock/SPL/PR. 172-173, J. Christopher Mihos and Lars Hernquist. 173 (t), Julian Baum/SPL/PR; (b), NASA, Hubble Space Telescope/European Space Agency. 174, Lockheed Research Laboratory, Palo Alto, in collaboration with the National Astronomical Observatory of Japan and Univ. of Tokyo, U. S. work supported by Marshall Space Flight Center of NASA, Japanese work by the Institute of Space and Astronautical Science. 175 (t), George Fowler/SPL/PR; (b), Roger H. Ressmeyer. 176-177, David Malin, Anglo-Australian Observatory. 178 (t), Roger H. Ressmeyer, Starlight; (b), Peter Menzel. 179-181 (all), Roger H. Ressmeyer, Starlight. 182 (both), NASA. 183, NASA/Hubble Space Telescope/digitally enhanced by Roger H. Ressmeyer. 184 (t), Roger H. Ressmeyer, Starlight; (b), Davis Meltzer. 185 (1), Christine Jones, William R. Forman, and Carolyn Stern/Harvard-Smithsonian Center for Astrophysics; (2), Lawrence Berkeley Laboratory, COBE Science Team/Differential Microwave Radiometer; (3), NASA/Goddard, COBE Science Team/Diffuse Infrared Background Experiment; (4), European Southern Observatory; (5), Richard C. Henry, Maryland Space Grant Consortium; (6), Joachim Trumper, Max Planck Institute for Extraterrestrial Physics; (7), Carl Fichtel, NASA/Goddard, Compton Observatory Egret Team. 186-189 (all), Roger H. Ressmeyer, Starlight.

**EPILOGUE:** 190, J. A. Kraulis/Masterfile. 192, Arthur Winfree/SPL/PR. 195, J. Bernholc et al, North Carolina State Univ./SPL/PR.

Library of Congress Œ data

Front line of discovery / (contributing authors, Arthur C. Clarke . . . et al.).
p. cm.
Includes index.
ISBN 0-87044-979-6
1. Discoveries in science. 2. Inventions. I. Clarke, Arthur Charles, 1917-
II. National Geographic Society (U.S.). Book Division.
Q180.55.D57F76 1994
600—dc20                                              94-38077
                                                          Œ

Composition for this book by the National Geographic Society Book Division with the assistance of the Typographic section of National Geographic Production Services, Pre-Press Division. Printed and bound by R. R. Donnelley & Sons, Willard, Ohio. Color separations by Digital Color Image, Cherry Hill, N.J.; Graphic Art Service, Inc., Nashville, Tenn.; Lanman Progressive Co., Washington, D. C.; Penn Colour Graphics, Inc., Huntingdon Valley, Pa.; and Prototype Color Graphics, Pennsauken, N.J. Dust jacket printed by Miken Systems, Inc., Cheektowaga, N.Y.